WE ARE THE TARGETS
The Story of Environmental Impact

WE ARE THE TARGETS

The Story of Environmental Impact

By

Harold J. McKenna, Jr.

Illustrated by

Nancy Lou Gahan

RICHARDS ROSEN PRESS, INC.

New York, New York 10010

Published in 1980 by Richards Rosen Press, Inc.
29 East 21st Street, New York, N.Y. 10010

FIRST EDITION

Library of Congress Cataloging in Publication Data

McKenna, Harold J
  We are the targets.

  (The Student scientist series)
  SUMMARY: Discusses the effects of various pollutants
on the environment and outlines what can be done to prevent
further environmental damage.
  1. Environmental impact analysis—Juvenile literature.
2. Pollution—Juvenile literature.
[1. Pollution]
I. Title.
TD194.6.M32   333.7   79–19827
ISBN 0–8239–0474–1

Manufactured in the United States of America

About the Author

Dr. Harold J. McKenna is an Associate Professor of Environmental Science in the Department of Secondary and Continuing Education, School of Education, The City College of New York. He holds an M.A. degree in Science Education from CCNY and an Ed.D. in Environmental Science Education from Columbia University. He is the Program Head of the Environmental Studies Graduate Program at City College; has taught graduate courses in ecology, conservation, field studies, and ethology, and has conducted research seminars. In addition, he is a supervisor of science student-teachers and has developed several science education programs that were funded by the National Science Foundation and the Department of Health, Education, and Welfare. Among them are an environmental science curriculum guide for secondary schools, a teacher-training program in environmental education, and most recently a health careers program for seventh graders.

Dr. McKenna is an active member of the New York Biology Teachers Association, NABT, AAAS, NSTA,

New York Academy of Sciences, SIPI, and various local environmental organizations. He is the author of several published articles on environmental science education and is presently co-authoring a text on environmental science for secondary school students and a book on career preparation for students considering a major in science.

Dr. McKenna is an adjunct professor of ecology at the New York Botanical Gardens and is presently serving as first vice-president of the newly established Weis Ecology Center in New Jersey.

Contents

WE ARE THE TARGETS
The Story of Environmental Impact

I

Introduction

Should the sale of cigarettes be outlawed? Can recycling paper really help save our trees? Is city water fit to drink? Is the use of pesticides in our foods causing cancer? Pollution of air and water isn't hurting me personally, so why should I worry about it? Isn't a little pollution the price we must pay for progress? Is it all that bad? We can't look back in time and return to the "good old days" of a crystal clear countryside as it was before we dropped our first candy wrapper. So what do environmentalists want? How do we know that we face an environmental crisis? Anyone not yet convinced that we do should immediately put down this book and look around. Go to the nearest river and decide whether you want to drink the water or even swim in it. Look at the sky and decide whether it is blue or brown. Check your grocery store and see what your foods contain.

In order to answer these questions and others about our environment, we must make decisions and learn to live with nature and its surroundings. We are simplifying our environmental systems and tampering with their checks and balances. What this all means is not clear to us, but we must give more thought to it if we wish to live in peace and harmony with all creatures large and small.

You as a consumer make decisions, and every decision you make has an environmental impact. That is an ecological fact of life! Every time you visit a department store, buy a ticket on an airline, or choose a place to live, your choices have an effect, for better or worse, on the quality of the water you drink, the air you breathe, the food you eat, the waste you generate, and most important the amount of resources you use.

One may look at environmental impacts as to cause and effect. For example, the two main *causes* of our environmental problems today are an ever-increasing population of human beings and a changing technology. If we look carefully at these two causes, we find that they affect the environment by having an impact on its resources. We then see the symptoms of these causes: air, water, and noise pollution; the increased amounts of solid waste generated; the energy crisis; and the overuse of pesticides to produce increased amounts of food to feed all the earth's people.

Too often we look at the symptoms of the environmental disease and try to cure them, only to find we still have the same problem. If we are to make an impact on our environmental problems, we must try to cure the causes of the disease. Once the causes are dealt with, the symptoms will lessen and eventually disappear.

In the chapters that follow you will come to understand the various factors that make the population and technology issues an environmental cause of the problem. We shall offer you some viable cures for these causes and explore the future of the problems. In addition, we shall analyze some of the symptoms of the problems and see how they can be managed and controlled in order to make the earth a healthier and happier place for human existence.

Before we look in detail at the various causes and symptoms that have impacts on our environment, we should examine a few basic laws that a leading environmentalist, Barry Commoner, has proposed.

The Laws of Ecology

1. *Everything needs everything else.* This first law shows how the earth and its living forces are interconnected. Through various cycles (chemical and biological) the earth is like a large scale, one side of which is balanced by what the other side does. For example, if a lake has a balance of oxygen, algae, fish, and other living things in it, it will remain stable as long as everything is in balance with everything else. By balanced we mean the proper number of organisms for the amount of food, space, and oxygen. If, however, there is an increase in the amount of nitrogen compounds, the algae will over take the lake through excessive growth. Ecologists call this eutrophication. With this increased growth the algae may choke out the fish, decrease the amount of oxygen, and cause the lake

to become unbalanced. It is therefore important that we understand how things in our environment are interconnected to other things if life is to continue to survive.

2. *Everything must go somewhere.* This second law emphasizes the fact that in nature there is no such thing as "waste." Too often we think that if we throw something away, it is out of sight and out of mind. We don't really care what happens to it. However, since we are in a "closed system" (on a small planet), all of our waste must go somewhere, since it does not leave the planet. A good example of this law is what happens to mercury batteries. When the battery is worn down, we throw it away. But where does it really go? First, we place it in a refuse can, then the garbage collector picks it up with all the other rubbish and takes it to an incinerator, where it is burned. Here the mercury is heated; this produces mercury vapor, which is emitted by the incinerator stack. This toxic vapor is carried by the wind and eventually is brought to earth in rain or snow. It can now enter a lake, and the mercury may condense and sink to the bottom. Here it is acted upon by microorganisms that change the mercury to another form called methyl mercury, which is soluble. This form of the mercury is taken up by fish and accumulates in their organs and flesh. The fish are caught and eaten, and the methyl mercury is deposited in human organs, where it might be harmful. In all, you can see that such waste ends up right back where it started, usually with man.

3. *There is no such thing as a free lunch.* This is a common-sense law in that it warns us that every gain is won at some cost, either to the environment or to us personally. Thus, we see the price we now must pay for our energy crisis, and in all the price that mankind will have to pay for polluting his environment. We cannot continue to plunder all of the earth's resources without paying some price. If we continue, the price we pay may have a far greater impact on our lives then we can presently imagine.

4. *Nature knows best.* This last of the laws of ecology is the one that most people find difficult to understand. Man too often tries to "improve on nature." For example, synthetics or man-made clothing are considered better than, let's say, cotton, a natural product. One must consider, however, the amount of energy, pollutants, and wastes generated by the manufacturing of synthetics such as nylon, as compared to cotton. Cotton is natural and as a plant carries out photosynthesis. No waste or additional energy is involved in the making of cotton. It is nature's way of doing things. On the other side, however, the production of the synthetic usually increases the amount of energy needed to manufacture it through use of fossil fuel, which also pollutes the air and water. In addition, harmful wastes may be generated. Another example is the man-made product DDT. If this product had been beneficial in nature, the chances are that there would have been a natural combination on the earth over millions of years to create it. However, man comes along and makes this poi-

son. He sprays it throughout the environment, causing all kinds of problems (this will be discussed in greater length in Chapter VIII). Again, nature carries on best for its own survival.

II

Impacts of Population Growth

As we saw in Chapter I, many factors contribute to man's impact on the environment, and one of the most important is *population*. As long as our numbers were few and our technology was simple, nature was able to handle any problems through the basic laws discussed previously. Hence we saw very few symptoms. They were basically diluted and taken care of through the system. In the last few centuries, however, technology has reduced the death rate and our numbers have been rising explosively.

In order to understand the impact of human population growth, it is important to gain insight into natural population growth. In nature most populations such as deer and wolves make a smooth transition of growth according to the carrying capacity of the environment. When the population size is small, the growth is exponential, but as the carrying capacity is approached the growth rate goes to zero and the population size remains constant. Such growth can be seen on a logistic curve or sigmoid curve (see Fig. 1). Here you can see how the growth of a population takes place over a given time period until some carrying capacity is reached, such as space or the amount of food available,

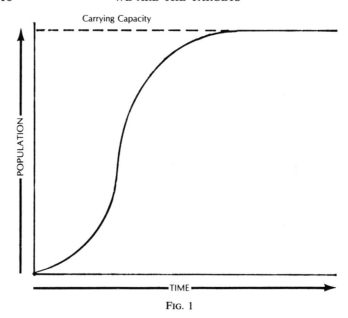

FIG. 1

and equilibrium of population growth is accomplished.

There are other ways of accomplishing equilibrium. If the population overshoots the carrying capacity, it will be forced to cut back until it again approaches equilibrium (see Fig. 2).

A more serious possibility occurs when the population overshoots the carrying capacity and in the process damages the environment to such an extent that the carrying capacity itself is decreased. Thus, the population is cut back to some value below the original carrying capacity (see Fig. 3). An example of Figure 3 occurs when cattle overgraze a meadow and damage the plant life so that it cannot recover.

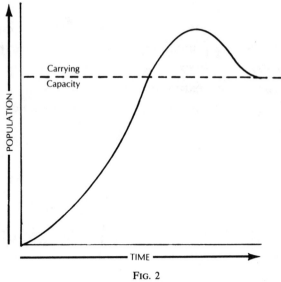

FIG. 2

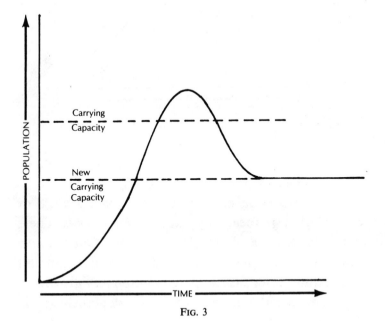

FIG. 3

The most extreme growth curve is the "J-shaped" curve, wherein a population sustains rapid growth and then drops off to an extremely low value at which point extermination or extinction is possible. Such a drop can be caused by outside factors, such as severe weather

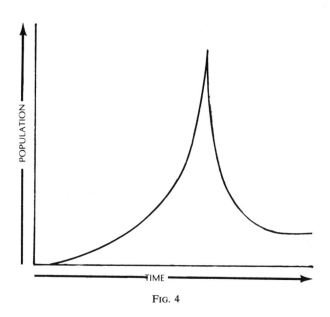

FIG. 4

conditions, severe pollution of the environment, or rapid decrease of food supply (see Fig. 4).

Before discussing growth rates in detail, a word or two on what is meant by *exponential growth*. There are two basic types of growth to consider. One is *linear growth*, in which something grows by a fixed amount in each time period, as compared to exponential growth, in which something grows by a fixed percentage in each time period. For example, if you invest $100 in something that earns 10 percent, your total investment will grow exponentially. However, if the $100 is not invested but each year $10 is added, then the total will grow linearly. In both cases the investment will be $110 after one year, but after 21 years the investment in the exponential example will be worth $800, whereas in the linear case it will be worth only $310.

Exponential growth is deceptively rapid. To test your understanding, try this: Ask your parents to give you a penny for the first day of the month, two cents for the next day, four cents for the next, eight cents for the next, and so on for every day of the month. Is this a better arrangement for an allowance than asking your parents for $5 per week? Why?

The natural increase in population depends upon the difference between the birth rate and death rate. If there are more births than deaths, population gains. If there are more deaths than births, population declines. To determine, therefore, the *growth rate* of a population we simply subtract the death rate from the birth rate. It is usually calculated as a percent of annual increase per thousand people. Thus, in the world in 1977, the birth rate was 34 per thousand and the death rate was 13 per thousand, which would give a growth rate of 2.1 percent. Of course, to be accurate one must also take into account the net migration rate of a population. You can see that world growth rate is just a little over 2 percent per year, and considering exponential growth you can readily see that the 1977 world growth rate of 2.1 percent per year corresponds to a doubling time of 35 years. The United States at a growth rate of 0.8 percent per year would double its population in 86 years. Table 1 summarizes some important population statistics for 1977, including crude birth rates, death rates, annual growth rates, and doubling time. In addition, a comparison of 1950 and 1973 is made.

By now you are probably asking yourself what all these statistics mean. The 2.1 percent world growth rate seems small until you realize that it means a net addition to the total world population of about 77 million people per year. To examine it even further, try to realize that with only a 2 percent growth rate we add about 2½ persons every second to the face of the earth! If we look back in history, we see that the world population growth rate was slower. Between the years 1000–1850 there were about 1 billion people on earth. The population doubled from ½ billion in 1650 to the one billion mark in 1800. It took about 120 years after this to double to 2 billion. Since the doubling time has not been constant, we know the growth has not been exponential, but it has been worse than that! The rate of growth has itself been growing.

What is causing this sudden rise in population growth rate? As we have seen, growth rate is the difference between the birth and death rates. A decrease in death rate causes sudden growth. People are living much longer these days. Even though the birth rate decreases, there are just too many people already on the earth.

Table 1. Summary of 1977 World Population Data

Area	Population Est. 1977 (millions)	Annual Birth Rate Per Thousand	Annual Death Rate Per Thousand	Growth Rate (%)	Doubling Time (years)	1950 Pop. (Mil.)	1973 Pop. (Mil.)	Population Estimates (Mil.) Year 2000
World	4,307	34	13	2.1	35	2,543	3,860	6,494
Africa	451	46	20	2.8	25	219	374	818
Asia	2,506	38	14	2.4	29	1,408	2,204	3,777
North America	247	14	9	0.8	87	166	233	333
Latin America	342	37	9	2.8	25	164	308	652
Europe	479	15	11	0.6	116	392	472	568
U.S.S.R.	259	18	9	1.0	69	180	250	330

Table 2. Determining Growth Rate

$$\text{Growth Rate} = \frac{\text{Birth Rate} - \text{Death Rate} + \text{Net Migration Rate}}{10}$$
$$(\%)$$

$$\text{Death Rate} = \frac{\#\ \text{deaths/year}}{\text{Population}} \times 1000$$

$$\text{Birth Rate} = \frac{\#\ \text{of live births/year}}{\text{Population}} \times 1000$$

$$\text{Net Migration Rate} = \frac{\text{Immigration} - \text{emigration}}{\text{Population}} \times 1000$$

Table 3. Summary of 1977 U.S. Population Data

Population Estimate in Millions (mid-1977)—224
Growth Rate—0.8%
Birth Rate—14
Death Rate—9
Doubling Time (Years)—87
Population Comparisons by Year
 1950—152 million
 1969—203 million

In some developing and underdeveloped countries, high growth rate is accompanied by large families, and the many social and religious restrictions placed on birth control and abortion all add up to increases in growth rate. To further compound the problem, the age structure of various societies in the world today shows a large percentage of women in childbearing ages (18–25), which usually means an increase in birth rate as well.

The interest and concern of scientists in population growth isn't new. In the 100 years between 1750 and 1850 the population of Europe nearly doubled. It was during that period that the economist Thomas Malthus said, in his famous essay of 1798, that the population growth at that time was so great that if it grew unchecked, it would increase much faster than the amount of food needed to feed the increased numbers. The disaster that he predicted never came about, since the three controlling factors in population were in operation. These factors, namely, *war, famine, and disease,* have now and in the past played an important part in controlling population growth. In the developed countries of the world, however, there are attempts to control these three factors through advanced technology. By controlling these factors man is in part also causing the sudden increase in population.

The human population, like other populations, cannot continue to grow indefinitely. When we ask which country is the most overpopulated, most people say China or India. In fact, the U.S. is probably the most overpopulated nation in the world. Why? We utilize more (nearly 50 percent) of the world's natural resources for a population that is about 10 percent of the world's. Of course, we have a high standard of living, but it is at the expense of other nations not able to keep up with us. As we already see, we are running out of fossil fuel needed to give us the energy to run our industry and homes; we are running out of other important resources that are used in combatting diseases and starvation. The explosive growth in the world's population will require rapid increases in food production if we are to survive even at inadequate levels of nutrition. The world is presently producing enough food to feed everyone adequately, but because of unequal distribution and waste in the world, about half of the people are suffering from hunger while others are overnourished. With the world population expected to double in 35 years, the prospects for widespread famine by the end of the century are extremely high. We already see parts of the world undernourished— that is, having insufficient amounts of food to provide energy (calories); and malnourished—not having sufficient amounts of protein in the diet.

Table 4. The Food Americans Eat Each Year

Food Item	1950	1977	Change
	(in pounds)		
Meat	125	155	Up 23%
Poultry	25	53	Up 113%
Fish	12	13	Up 9%
Eggs (doz.)	31	22	Down 28%
Milk, cream (gal.)	40	34	Down 14%
Cheese	8	16	Up 106%
Butter	11	4	Down 59%
Sugar	100	95	Down 6%
Flour	135	110	Down 18%
Soft drinks			
(12-oz. can)	100	325	Up 225%
Vegetables (fresh)	115	102	Down 11%
" (processed)	84	123	Up 46%

To meet the increasing demand for food as the population increases will place a great deal of stress on the environment. It means that we must:

(1) *Change the eating habits of people.* Refer to Table 4 to see some of the foods people eat. As you can see, people in the U.S. are eating more food than before, but these foods are not as efficient in energy as others. We would be far more efficient and gain more energy from our food if we ate plants rather than meat. Green

plants are low in the food chain and have more energy available to consumers. There is a tremendous energy loss through heat energy and respiration because we are meat eaters or carnivores. If we look at the proteins, we find we can raise high-protein soybeans, peas, and beans utilizing less land than raising beef cattle and hogs. Soybeans could be a much more important source of protein in the future if they were used directly as food for humans instead of as feed for animals. These plants are extremely efficient producers of protein. Among animals, fowl are more efficient than hogs, which in turn are more efficient than beef cattle. Yet we find that beef consumption has increased by 90 percent since 1950.

It is important to note that animals are inefficient sources of food when and only when they are consuming food that we could consume directly ourselves. When animals consume wastes or graze on plant products not directly used by man, they become an extremely important supply of food to man, since there is no competition.

(2) *Look to the oceans for the answer.* In examining the world's consumption rate, we find that fish supplies only about 0.8 percent of the calories and about 4.5 percent of the protein, so that it is presently a minor food source. Looking toward the future, many scientists feel the oceans will not serve as a source of food in the quantities needed since most of the fish live in the coastal waters, where the highest amount of pollution is found. Studies have been made on the possibility

of grinding fish into fishmeal and then feeding it to cattle. In this way, we would not utilize the land resources that man might use directly for growing food items. The harvesting of large amounts of plant life from the oceans is being considered to a degree, but we are not sure of the nutritional value of various species or their effects on us. Some sea plants have even been considered poisonous to man. We therefore need more research into oceans and their future use, but we still need to utilize our land resources.

(3) *Use the land more efficiently.* Since it appears that the oceans have rather limited use as a food source, it will be necessary to increase food production on the land to keep up with the increasing population. To do this we must increase the yield per acre and the acreage being cultivated. The total world land area is about 32 billion acres. Of this, 16 billion acres are mountains and deserts and not suitable for agriculture. Of course, some of these acres will be used through converting the deserts into usable land at the cost of changing the desert environment. This in turn may affect other environments. We just don't know all the interconnections and impacts. There are 8 billion acres of arable land, of which half is already farmed. Another 8 billion acres is grazable land, of which 50 percent is already being grazed upon. A good diet (2,000 kilocalories of high-quality protein daily) requires 1 acre of arable land and 1 acre of grazable land per person. Thus, if there is a total of 16 billion acres in the world available, it can only support a total of 8 billion people.

If we continue growing, we will outgrow our available land resource.

To increase the yield of the land requires progress in plant genetics, mechanization, irrigation, artificial fertilizers, and extensive use of pesticides. In this area, we need to develop disease-resistant plants, drought-resistant strains of plants that can be grown in different parts of the world throughout the various seasons. In addition, we are becoming dependent on advanced technology that in turn unbalances the natural cycles found in the various environments. One must remember that developing artificial fertilizers is taking far more energy and tipping the scale at one end to satisfy another. This is not the answer to the problem.

If we can accept the fact that the world population is increasing and recognize the problems involved in feeding a hungry world, then we must turn to the inevitable—how can we keep the world from becoming over-populated in terms of the resources available? This is a difficult question to face, since it involves moral judgments of highly controversial issues. If we wish to stop growth we must simply increase the death rate or decrease the birth rate—or even better, do both. Obviously, the easier alternative to tackle is that of controlling the number of births. Since war, famine, and disease are no longer forces great enough to control the numbers we are reaching, we must look for man-made controls. In nature, populations control their numbers in various ways. They compete for food and space; they move to other areas where there is less competition; or they increase their numbers to a point where starvation, aggression, and disease take over. In this way, nature insures the survival of the fittest and that one carries on. In man we are controlling the "natural way" through our advanced technology, and therefore we need to control our numbers through our own efforts. Four basic methods of controlling our births are:

(a) Contraception (IUD, condom, pill, diaphragm, spermicides, rhythm). This method interferes with the fertilization of the egg, thus decreasing the chances of a birth.

(b) Sterilization (vasectomy, salpingectomy). This method is permanent, and insures there will be no fertilization of the egg at any time.

(c) Abortion. This method insures that a birth will not take place, since an already fertilized egg and developing embryo are removed.

(d) Family planning programs. Family planning programs make contraception information and services available to help individuals determine their own family size. In this way millions of motivated couples avoid unwanted pregnancies. The Population Council reported that, in 1970, 23 developing countries had such programs and 15 provided some support without an explicit policy. Sixty-one developing countries have none at all.

If as a society we are ready to use one or more of these methods, we then may decrease the rate of births each year. This, in time, may give us the wisdom and foresight to explore further more advanced techniques of other population controls in order that there be a good quality of life for all people on this small, closed environment that we call the earth.

III

Impacts of Technology

We have seen in the preceding chapter the impacts that population growth can have on the world's environment, making it one of the major causes of environmental deterioration in the world today. Now, let us look at the second major cause of the environmental crisis in the developed world today—that of a changing and ever-growing *technology.* It is found that whereas production for the most basic needs—clothing, food, housing—has just about kept up with the 50 percent or so increase in population, the kinds of goods produced to meet these needs have greatly changed. New production technologies have taken over the old ones. For example, synthetic detergents have replaced soap powder; nylon and dacron have replaced cotton and wool; aluminum and plastic have replaced steel and lumber; truck freight has replaced railroad freight; and nonreturnable bottles have replaced returnable ones. In addition, older methods of pest control have been replaced by newer ones, such as chemical pesticides (DDT, Diazone); and for controlling garden weeds the cultivator has been replaced by the herbicide spray. Range feeding of cattle has been displaced by feedlots.

The important fact in each of the above cases is that the technology itself has changed drastically rather than the overall output of the economic good. Basically the individual need has remained the same in terms of consumption of food, use of clothes, occupation of housing. Today, however, our food is grown on less land with more fertilizer and pesticides than in the past; our clothes are made of synthetic fibers rather than natural ones; we use synthetic detergents in laundering rather than soap; we live in buildings that depend more on aluminum, cement, and plastics than on steel and lumber; our goods are shipped by truck rather than rail; and we drink our soda out of nonreturnable bottles or cans rather than out of returnable ones.

This pattern of economic growth and the changes in technological production are the major reasons for some of the symptoms we see in our environment. Such symptoms as increased solid waste, pollution of air and water, use of synthetic organic chemicals, and problems in increased energy production are all part of the environment scene that we are facing today and can face in the future.

Let's take a closer look at some specific cases. Start with agriculture. Some of the technological developments that have taken place are heavy farm machinery, feedlots, inorganic fertilizers, and synthetic pesticides. Much of this new technology has been an ecological disaster. Take, for example, the feedlots. In such lots cattle spend most of their lives being fattened for market. They are fed with grain, rather than being allowed to graze on pasture lands. As a result, most of the Midwest has converted good pasturage land into intensive grain-producing land to meet the increased need. In addition, the humus content of the soil is depleted, and farmers are forced to use inorganic fertilizers to get their high yields of grain. This unbalances the ecological sequence of the natural pasture land, which did not require the use of artificial fertilizers since the plant life growing in the pasture was adapted to that environment. One can further see that each year farmers are required to use more artificial nitrogen fertilizers to increase their crop yield. This figure, however, can be deceiving, since the amount of increased fertilizer applied to the soil is far greater than the crop yield produced. We are actually overfertilizing the soil. This is so because much of the fertilizer is not taken up by the plants. Instead, it runs off as waste into the environment, causing an imbalance. This new technology is an economic success but an ecological disaster. Furthermore, in the feedlots there is the problem of cattle waste. It remains in large quantities in the feedlot area, where it is leached into groundwater beneath the soil or may even run off the slope of the land into various waterways. This heavy concentration of animal waste may cause problems with the waterways by "overpolluting" them. In such a situation, the water becomes overenriched with chemicals found in the waste products (nitrates), causing an unbalancing of the plant and animal communities living in the body of water.

A second case is that of pesticides. We have produced more synthetic insecticides within the past fifty years than ever before. It is the old story of higher crop yields, more pesticide used, less efficiency, and increasing environmental problems. The main problem of using chemical pesticides, such as insecticides, is that the insecticide is sprayed onto the crops, killing off the pest as well as the natural pest predator. As the insect pest builds up a resistance to the chemical sprayed, we must increase the amount used to counterattack the adaptation of the insect. So you see this whole cycle of increased use to get increased yield is increasing the impact on the environment.

A third case of technological causes for many of our environmental problems is that of the use of detergents in laundry instead of soap. Basically, the detergent has replaced the natural organic product of soap by an unnatural synthetic one. Soap is produced by combining a natural product, such as fat or oil, with an alkali. The fat product used in soap-making usually comes from a plant, which is natural and uses CO_2 and H_2O as the raw materials. Sunlight provides the energy to carry out the photosynthetic process. All of these materials are available in nature free of charge and are renewable resources. It should be kept in mind, however, that if conservational practices are not used in the care of the plants producing the fat or oil, it can cause soil problems and also affect the environment. It is also important to note that in the manufacture of soap, energy is used in the form of fuel consumption and waste products are produced.

Soap, once used and sent down the drain, is usually broken down by microorganisms such as bacteria. The products released from this process are CO_2 and H_2O. This is so because fat contains only carbon, hydrogen, and oxygen atoms. Thus, there is very little impact on the aquatic environment since these compounds are readily found in nature.

In contrast, however, the production of detergent has more intense environmental impact. Detergents are produced from organic raw materials that are found in petroleum along with other materials. In order to obtain these materials, petroleum is distilled. This takes a great deal of energy, and in the process the burned fuel may pollute the air. These distilled materials are further synthesized through a series of chemical reactions involving chlorine and high temperatures. This in turn gives us an active cleaning agent that not only gets our clothes white, but "whiter than white." The total energy used to produce this active cleaning agent is about three times that needed to produce oil for soap-making. In addition, until recently detergents would not break down as soap did by the action of bacteria. They therefore remained in the aquatic environment as mounds of foam. Fortunately, under new legislation, detergents now must be "biodegradable,"— able to be broken down. The problem with this, however, is that when the detergent is broken down it releases a toxic material that can kill fish. So we still have the problem. Thus far technology has either not been concerned or has found no solution to it. Of course, this is only the beginning of the story. We can add to it the phosphate factor found in detergents that is in part causing pollution of our waterways, and the matter of packaging the detergent for sale.

Let's take a closer look at the technological displacement of our rubbish, most of it from packaging. Take, for example, the nonreturnable bottles or cans. When we throw them away in our rubbish container, they are unable to be naturally cycled in the environment. Thus they accumulate and have to be collected and "recycled" at a tremendous expenditure of energy and pollution production or remain as an ugly pile of our technological advance. In the days before the nonreturnable bottles, people used to return their bottles to the store, where they received a return of their deposit. The storekeeper then would ship the bottles to the company, where the bottles would be washed, sterilized, and reused. Usually a bottle could be reused about four times. Thus, we were saving energy in this process. By not returning the bottles, but simply throwing them away, we need to break the bottles and recycle the glass fragments into new bottles, at a tremendous cost in energy.

A similiar situation exists with cans. It takes about six times as much energy to produce an aluminum can as a steel can. Yet we are using more and more aluminum for this purpose. Furthermore, in our packaging of foods, we are adding a layer of plastic as an inside lining. It has no human value, but it does increase the problem of nondegradable plastics in the environment that is also mounting.

At this point, you should have some insight into the technological cause of environmental problems. It is now time to look ahead and see how technology can reshape our lives in the future. Recent discoveries could improve widespread use of the sun as an energy source; fiber optics may alter our communications system by sending messages on a beam of light; the family car may develop using new and different fuels; and the computer that can read, write, talk, and understand the spoken word may do complex chores for us, enabling people to have more leisure time. Of course, all of these new ideas have to be examined in terms of the impact they may have on the environment. In any event, a tremendous amount of money is being spent on technological research in order to combat some of the existing crises that are facing developed nations today. As environmentalists, we must carefully look at these new developments and determine their impacts, and make the public aware of both the pros and cons of such technologies. A few new developments that are being worked on today are the following:

Energy

A major achievement in this area are solar cells. These solar cells or photovoltaic devices convert the sun's rays directly into electricity. They are made of finely sliced silicon crystals. These silicon crystals are very expensive and inefficient, and to date have not' been practical to use in electric power generation. Several new ideas are developing in this area. One is the use of amorphous silicon, which is less expensive to produce, and which can be sprayed on a conducting surface, not cut like the crystals. In this way cost is reduced, and thin layers can be used on the surfaces. Furthermore, it has been found that gallium arsenide is more efficient in converting the sun's radiation to electricity and in withstanding high temperatures that silicon cannot. With the use of concentrators (similar to large magnifying glasses), sunlight can boost efficiency without driving up the cost. This could provide electricity more efficiently without burning fossil fuel to generate steam as we have been doing. Instead, we can use our technology and the sun to give us the increased production of electricity.

The magnetohydrodynamic technology (MHD) is one of the most promising means of providing electricity that someday may run millions of homes. In the MHD process, gas from any fossil fuel is heated to about 5000°F, then sent through a magnetic field to create electrical current. In this way, MHD eliminates the mechanical steps of converting fuel to steam by burning fuel. It improves efficiency by 50 percent, which means less waste and lower cost to the consumer. Demonstration plants in the U.S. are planned in 1990.

Computer Technology

Computers already are important in U.S. households as part of the newest ovens, dishwashers, and TV sets. Their popularity is increasing every year. In addition to these, inexpensive computers for the home are on the way. Such computers might pay bills, keep financial records, plan meals, keep food inventories, and even order groceries. We are going to become more and more reliant on these machines in the future, and perhaps they will be the important link between our technology and the understanding of our environment.

Light Wave Technology

Engineers have learned how to use light to send messages at higher speeds and greater volume than anything possible with electricity and copper wire. Phototronics, as it is called, could become as common as electronics. Some day we may be able to substitute light for electricity as an energy force. This, of course, means rethinking of fundamentals, and here lies the challenge to both science and technology.

At present, light-wave technology is in its infancy, but it is being developed in degrees by the Bell System for use in the telephone. How does it work? A miniature laser beam converts voice signals on a telephone from electrical impulses to light. The laser beam, blinking on and off millions of times a second, is aimed into one end of a glass fiber as fine as a strand of human hair. The unique property of glass allows the light going through it to be sent for miles, losing little strength. A bundle of 200 fibers, no larger than a cigar, can handle about 67,000 calls at once. In addition. it can carry video transmission, such as television, which the regular copper phone lines cannot. You can see, therefore, that not only can your voice be carried through these optical fibers, but TV and even computer linkups might be all sent at one time in the future.

Transportation Technology

Researchers assert that the internal combustion engine will power America's automobiles well into the 21st century. What probably will change is the fuel. As the price of gasoline increases, more cars will use synthetic fuels made from coal or even green plants, such as corn. Alcohol can be made from corn that may be used as a fuel for cars and trucks. It has been found recently that a combination of gas and alcohol, Gasohol, boosts fuel economy and octane rating and is comparable to unleaded gas in its low pollution effect. It is hoped that foreseeable advances in automobile technology might make the car freer of pollution and less of an energy drain. However, the automobile technology will still not eliminate rush-hour traffic jams or highway fatalities. These are dimensions that are not in the total realm of technology. The human element is still important in overcoming many of our social problems, of which the environmental concern on how things have impact on other things is still important.

Let us now analyze some of the symptoms in our environment that have been caused by too many people and a rapidly changing technology.

IV

Solid Waste Pollution

Of all living things, only man can consciously manipulate, control, destroy, or preserve his environment. One of the symptoms of man's manipulation and control of his environment is the tremendous amounts of solid waste that are generated at an ever-increasing rate. The United States needs to assess the impacts that solid waste are having on the environment and develop a program with specific objectives to deal with the problem. Some of the objectives that should be considered are to:

1. Minimize the cost and maximize the effectiveness of collection, treatment, and disposal of solid waste.
2. Minimize the amount of solid waste generated per unit time.
3. Maximize the percentage of solid waste that can be separated and reused economically.
4. Minimize the negative impact of solid waste disposal on all ecological systems.

Before looking closer at programs in which these

In an industrial community, a street is cluttered with refuse.

objectives are dealt with, let's first see what we mean by solid waste. Solid waste can be classified into several types. First, *garbage* is waste that comes from growing, preparing, cooking, and serving food. Thus, orange peels, bones, and so forth are considered garbage. Second, *rubbish* consists of combustible items such as boxes, grass, and clothing, and noncombustible items such as cans and glass. Finally, *refuse* is a general term used to encompass all types of waste.

Part of the problem is people's attitudes, affluence, and methods of collection and disposal. For example, the U.S. generates about 3.5 billion tons per year of refuse. We are basically "waste makers." Of this figure, about 200 million tons per year is household waste, not from industry. It is figured that about 8 pounds per person per day will be the output of refuse by people in the U.S. by 1980.

The problem of solid waste is one that we have dealt with in the past. We have an "open cycle" of waste. That is, our economy dictates production. In turn, production involves distribution and marketing of products that all go into use and discard. Discard involves collection and disposal. Once our refuse is out of sight, it is out of our minds. But we still have the problem of how to dispose of increased amounts of refuse over a short period of time (see Fig. 5).

In dealing with the refuse problem we must understand how it is collected and then disposed of. The usual methods of disposal have been open dumping, incineration, and sanitary landfill. Until recently, very

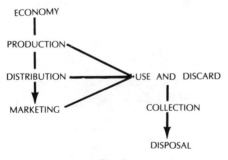

FIG. 5

little has been done in disposing of wastes in a pollution-free way or reusing them efficiently.

In open dumping, microorganisms eventually attack biodegradable garbage and decompose it. Unfortunately, today there is too much garbage for the amount of space available, and it can become a health problem. Furthermore, unless we separate our refuse into degradable and nondegradable, this procedure can only add to the problem.

Incineration is a good method in that it reduces the volume of waste, but we still have to deal with the by-products produced, namely, ashes. In addition, we must deal with pollutants that can enter the air through the process.

Landfill techniques are presently being used in which refuse is dumped in various ways. One such method is that of area fill, in which refuse is deposited in horizontal layers on flat ground, compacted, and covered over with soil. As a result layers are built up, and decay of organics eventually decomposes the materials. Another method is that of trench fill, in which the land is excavated in the shape of a trench to a depth, length, and width that is needed to accommodate the volume of waste being dumped. The trench is then filled in with soil and compacted. This method does not have the layering effect as in the area method. Instead, the landfill area is kept at the same base level as the land.

With diminishing land within a reasonable distance from our major cities where the largest quantity of

Garbage collection in a large city is a highly organized activity.

Landfill is one method of disposing of large quantities of refuse.

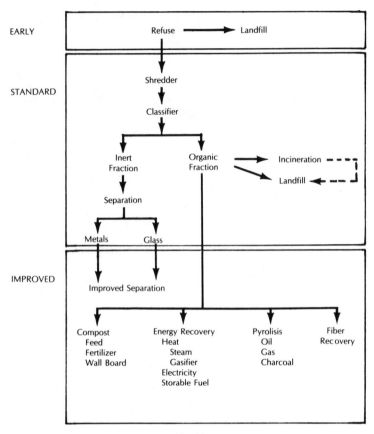

FIG. 6

waste is generated, and with our concern about pollution and the need to conserve dwindling natural resources, efforts are currently being made to find some answers to the refuse problem. One way to deal more effectively with solid wastes is that of a *materials recovery system*. (see Fig. 6).

In this operation, mixed refuse is first shredded for
size reduction; an air classifier then separates the light
and heavy fractions; iron is magnetically separated out;
liquid media tanks separate the larger pieces of glass
and aluminum from other non-iron metals. The mixed
aluminum and glass then is separated electrostatically.
Finally, the remaining glass is optically sorted into
clear, green, and amber colors.

As shown in Table 5, many valuable materials can
be recovered from solid waste. We just need to develop
the system that can do the job to recover these needed
resources.

We are now working on methods of converting organic refuse into new energy-yielding resources. In this
way, we may also meet the challenges of the shortages
of fossil fuel, and use the waste materials to benefit
mankind. For example, in St. Louis, Missouri, the
Union Electric Company burns about 300 tons of the
city's refuse each day in modified coal-burning boilers.
This makes up about 15 percent of the fuel requirements
of the city. We further are developing other forms of
incineration, such as *pyrolysis.* In this process, solid
wastes are baked in closed chambers at very high temperatures in the absence of oxygen. This reduces the
volume of the waste by about 90 percent. The organic
wastes are decomposed into acids, alcohols, and gases.
It has been estimated that one ton of pyrolyzed solid
waste yields 18,000 cubic feet of gas; 114 gallons of
liquid, which is about 90 percent water; 25 pounds
of ammonium sulfate, which can be used as fertilizer;

Table 5. *Value of Materials in Solid Waste*

Resource	Tonnage Quantity (× 1000)	Typical Price ($/Ton)	Total Revenue Potential (× 1000)
Ferrous Metals			
1. Containers and closures	5,400		
2. Domestic and commercial durables	3,500		
Total	8,900	$ 15	$133,500
Paper			
1. Newsprint	7,600	10	76,000
2. Corrugated	10,000	10	100,000
3. Other	20,000	1	20,000
Total	37,600	$ 21	$196,000
Glass			
1. Containers	9,700		
2. Other	1,100		
Total	10,800	$ 12	$129,600
Aluminum			
1. Cans and closures	250		
2. Foil	200		
3. Consumer durables	180		
Total	630	$ 200	$126,000
Tin			
—total production			
1. Tin plate	28		
2. Foil	1		
3. Other	37		
Total	66	$2,000	$132,000
Copper			
—total production	355	$ 600	$213,000
Lead	30	$ 80	$ 2,400
Textiles			
—total production	2,900	$ 10	$ 29,000
Rubber			
—total production	1,600	$ 5	$ 8,000
Plastics			
1. Containers and film	1,900		
2. Toys	200		
3. Consumer appliances and durables	100		
Total	2,200	$ 10.50	$ 23,100
TOTAL			$992,600

a half gallon of tar; and about 154 pounds of solid residue of metals, glass, and other materials that can be recovered and sold.

Thus far we have discussed only one aspect of the solid waste problem, that of refuse. There is still the

Man and machine team together to cope with mounting piles of trash in the streets.

problem of municipal waste and sewage treatment. With increases in such waste we should be building more and more treatment plants.

Sewage is the water supply of the community that has been polluted by use in the home and industry. Until recently, the disposal of waste water was not considered a problem. Sewage was merely allowed to

RICHMOND, Va.—Conventional in appearance, but made from unconventional materials, this home in a Richmond suburb is the first built almost entirely of recycled materials. Recycled aluminum beverage cans, glass bottles, newspapers and other waste products are used in building products that are either available now or are technically practical.

46

Labels on drawing:

SHEATHING OF RECYCLED PAPER

PANELING OF RECYCLED PAPER, VINYL AND BURLAP ADHERED TO ALUMINUM

PATIO DOOR FRAMES OF RECYCLED ALUMINUM

BARRIER BLOCK WITH CRUSHED GLASS

DUCTWORK OF RECYCLED ALUMINUM

CARPETING OF RECYCLED VINYL FIBER. CARPET PADDING OF RECYCLED JUTE

ROOF DECK OF RECYCLED PAPER

FLOOR TILE OF VINYL WITH ALUMINA TRIHYDRATE

ASPHALT ROOF AND WALL FILL OF RECYCLED PAPER FIBER FROM RECYCLED NEWSPRINT

CABINETS OF NON-COMMERCIAL SOUND WOOD, WOOD SCRAPS AND SAWDUST

INSULATION OF RECYCLED GLASS AND VINYL MILL

SUBFLOORING OF RECYCLED PAPER

WATER PIPE OF RECYCLED SCRAP COPPER

FLOOR JOISTS OF RECYCLED PAPER

ASPHALT ROOF SHINGLES OF RECYCLED TIMBER AND FIBER

ROOF TRUSSES OF RECYCLED ALUMINUM

FRAMING AND STUDS OF RECYCLED ALUMINUM

INTERIOR DOOR FRAMES OF RECYCLED ALUMINUM

SANITARY FIXTURES OF MARBLE CHIPS AND QUARRY TAILINGS

FLOOR TILE OF VINYL SCRAP AND WOOD SCRAP

SEWER AND VENT PIPES OF RECYCLED CAST IRON

FLY CONCRETE SLAB CONTAINING FLY ASH

CONCRETE FOOTING CONTAINING FLY ASH

FASCIA SOFFIT AND RAIN CARRIERS OF RECYCLED ALUMINUM

WINDOW FRAMES OF RECYCLED ALUMINUM

SIDING OF RECYCLED ALUMINUM

BRICK OF CRUSHED GLASS AND QUARRY TAILINGS

DRIVEWAY OF CRUSHED GLASS, SWALLOWLD RUBBER TIRES

LAWN CONTAINS COMPOST OF PROCESSED GARBAGE

CRUSHED GLASS FILL

PHOTO FROM: REYNOLDS METALS COMPANY, PUBLIC RELATIONS, RICHMOND, VA.

RICHMOND, Va.—This cutaway drawing pinpoints the location of the various recycled materials used in an apparently conventional suburban home built here by Reynolds Metals Company. The first home built almost entirely of recycled building products, it is priced competitively with similar homes built from conventional materials.

47

flow into the nearest body of water, thus solving the problem, it was thought, by dilution. However, with the rapidly expanding population and a sharp increase in industrial wastes, the hazards of this practice are readily seen. Many parts of the U.S. are familiar with contaminated drinking water, the loss of marine life and waterfowl, condemned bathing beaches, and the unsightly appearance of beautiful waterways.

It is now agreed that sewage control is essential, and many systems of sewage treatment have been devised. These include septic tanks, filter beds, and more sophisticated treatment plants.

Septic tanks are widely used for homes in areas where the population is sparse and where there is sufficient land for drainage of the tank. This system is used when it is not practical to have a central sewer system. The choice of a sewage treatment plant is determined by local conditions, such as volume of sewage, availability of land, and the nature of the receiving waters in which the by-product from the plant is to be discharged. Figure 6 shows a typical sewage treatment plant.

The treatment process begins when the raw sewage from homes and factories (influent) flows through the *bar screens.* These screens consist of two series of upright bars. Large objects, such as rags and sticks, are removed by the automatic raking devices. In addition, large amounts of grit, sand, and gravel enter the sewers, especially during rainstorms. It is necessary to remove this grit soon after it enters the plant in order to protect the equipment from abrasion and prevent a buildup of such materials. From the bar screens, the sewage

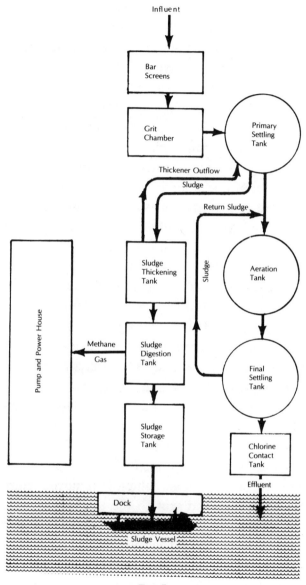

FIG. 7

travels to the *grit chamber.* This is a tank designed to slow the velocity of the sewage to about one foot per second. At this slow speed the heavier particles settle and are removed from the tank by a scraping mechanism. The lighter organic impurities remain suspended in the sewage flow for additional treatment. The above steps in the process are all preliminary; now primary treatment can begin.

The sewage during primary treatment is allowed to settle for about one hour in the *primary settling tank,* where about one-third of the suspended organic impurities settle and form a sludge. This sludge is removed by pumps. Small amounts of grease rise to the top and are removed by scraping devices.

Secondary treatment begins in the *aeration tank,* where sewage is stored for about six hours, depending on the process employed. During this time the sewage is supplied with compressed air to keep the materials aerated for bacterial action. The bacteria convert the dissolved and suspended solids into particles that will settle quickly. There is thus an accumulation of bacterial clusters called floc, which permits good settling. The aeration tank is followed by the *final settling tank,* where most of the suspended solids are now readily settled and form a sludge, which is removed from the bottom by pumps and sent to the thickening tank.

At the *thickener* the sludge further settles, and large amounts of water are separated from it. This water is returned to the sewage flow at the beginning of the plant and the remaining concentrated sludge or thickened sludge is pumped into digestes. The *digester* is a large enclosed circular tank in which the sludge is stored for several weeks at a temperature of about 95° F. These conditions favor the growth of anaerobic bacteria, which thrive without oxygen. These bacteria obtain their nourishment by breaking down the organic

The modern sewage treatment plant is a technological wonder.

compounds and in this way convert about half of the organic matter in the sludge into various gases and water. The gases mainly are methane and CO_2, and many treatment plants use methane gas to generate all or most of their power requirements. From the digestion tank, the digested sludge is pumped into a sludge *storage tank,* where it waits to be carried by ships out to sea and dumped.

The liquid wastes that have been processed and are present in the final settling tank are passed through *chlorine contact tanks* to kill any dangerous bacteria before being permitted to enter the waterways as *effluent.*

As can be seen, under such treatment the hazardous impact is somewhat lessened but still not eliminated. We need to be able to remove all organics and inorganics from waste products. In the future we may see treatment plants developed at the *tertiary level:* advanced treatment of sewage to remove all organics and inorganics. This can be accomplished through chemical oxidation, coagulation, adsorption, and desalinization. It is, however, an expensive operation.

Let's look at some more recent solutions that have been proposed. First, we need additional voluntary reclamation programs, involving collection of containers by individuals, their return to reclamation centers, and usually some payment for doing so. Reclamation programs for aluminum containers have been the most successful because of the relatively high salvage value. If other containers could have similar values as our resources decrease, perhaps we might have more of these programs.

Second, we need mandatory reclamation programs. In such programs deposits are required on all containers, and they provide for return of the deposit on return of the container to any retail store. In this way we can reuse valuable containers over and over again.

The third suggested program that has come to attention is that of taxing refuse by its weight. In other words, we pay for the waste we generate. The more waste we have in weight, the more tax we pay. In this way, people might think twice about waste, reuse their grocery store bags to carry more groceries, and reuse much of the packaging that is now thrown out. Furthermore, industry might react to this by developing more efficient packaging for products, such as a cube-shaped package that can contain the same volume but with less waste in the packaging.

V

Impacts of Air Pollution

Air pollution has been known to kill. In New York City, in London, England, and in Pennsylvania, polluted air that remained for several days caused serious illness and death, especially among the elderly and infants. Air pollution can also impair health. Dirty air makes eyes water and smart; it stings the throat and upsets the breathing. People with chronic lung or heart disease are very vulnerable to air pollution. Studies have shown direct relationships between prolonged exposure to polluted air and the incidence of bronchitis, asthma, emphysema, and cancer.

Before discussing specific pollutants found in the air and their effects on the environment, let's see what we mean by the word *pollution.* Pollution is the unwanted side effects of human activity. It is a symptom of the overpopulated earth and of man's changing technology. One way of describing a pollutant is to say that it is a *misplaced resource.* In this way we can realize that only resources that are out of place are considered pollutants of the environment. For example, CO_2 is a very valuable gas that is present in the atmosphere. It is needed for life to continue, because it is one of the basic chemicals used to make food in plants through photosynthesis. However, if CO_2 is "misplaced," or put where it does not belong or in too large a quantity, it can be considered a pollutant. Why? One main reason is that it can harm humans and other living things. When too abundant it heats the earth and causes problems as a result of the increase in temperature. What would be some other reasons?

When reading about various pollutants, you may have noticed the term *ppm,* or parts per million. This is the measure that is used. But what does it mean? Take for example the gas neon, found in the atmosphere to the extent of about .0018 percent. To find out the parts per million of this gas we simply develop a proportion such as:

$$\frac{.0018}{100} = \frac{X}{1,000,000}$$

In this way, we find that there are 18 ppm of neon in the atmosphere. Another way of looking at it is as follows: 1 ppm is equal to 1 inch in 16 miles, or 1 minute in 2 years. As you can see, there are many ways of figuring a part per million.

Pollutants in the air are usually classified into three types: Inorganic gases, organic gases, and particulates.

Inorganic Gases

Oxides of Sulfur. These are found in fuel combustion, burning of trash, chemical plants, and metal processing.

Sulfur is a nonmetallic element found in fuel oil and coal. When these fuels are burned, sulfur combines with oxygen in the air to form gaseous oxides such as sulfur dioxide (SO_2) and sulfur trioxide (SO_3). The sulfur oxides, in combination with water and oxygen from the air, can develop sulfuric acid (H_2SO_4), which can eat away iron gates, marble sculptures, and plant tissues and even affect man's respiratory system.

Oxides of Carbon. These include carbon dioxide (CO_2) and carbon monoxide (CO). In particular, CO comes from the internal combustion engine of the automobile and from solid waste disposal plants. Its chemistry is that of an invisible, odorless, tasteless gas formed when any carbon-containing fuel is not completely burned to CO_2. Its effects are readily seen in living animals: CO combines with the hemoglobin in the red blood cells instead of the needed oxygen; this in turn causes oxygen deprivation and can lead eventually to the death of tissues, cells, and the organism.

Oxides of Nitrogen. These include nitric oxide (NO) and nitrogen dioxide (NO_2). They are formed mainly through fuel combustion and in chemical plants. Because nitrogen makes up about 78 percent of the air around us, it becomes a pollutant when it combines with oxygen from the air to form the various oxides. We have recently found that it affects vegetation, destroying tissues and reducing growth. We also find that oxides of nitrogen cause dyes in fabrics to fade and even damage the fabric fibers. The nitrate salts, formed from the oxides, cause metals to corrode.

Organic Gases

Hydrocarbons. Methane and butane are usually formed by internal combustion engines, evaporation from painting and dry cleaning, agricultural burning, and the marketing of gasoline. They cause plant damage and respiratory irritation. The main problem with hydrocarbons is their role in photochemical oxidants. This is commonly known as photochemical smog. It causes eye irritation and weakens other materials.

Aldehydes and Ketones. These comprise formaldehydes and acetone, which are produced by chemical plants. They also affect the environment by being able to alter growth and even to cause death in living things. In addition, the chlorinated hydrocarbons that make up many insecticides are grouped as pollutants (see Chapter VIII).

Particulate Matter

Particulates exist as solids in the form of dust and smoke, and as liquids in the form of mists and sprays. This type of pollution results from industry and agricultural operations and from the exhaust of automobiles through their combustion. The effects of such particulates can be seen in terms of health—lungs pick up smoke and dust; particulates suspended in air absorb sunlight and reduce visibility and the amount of energy reaching the earth from the sun.

According to the Environmental Protection Agency

(EPA), the national emissions of major air pollutants in 1969 were as follows:

NATIONAL EMISSIONS OF MAJOR AIR POLLUTANTS: 1969
(latest year for which figures are available)
(millions of tons per year)

Source	Sulfur Oxides	Particulate Matter	Carbon Monoxide	Hydro- carbons	Nitrogen Oxides	Total
Transportation	1.1	0.8	111.5	19.8	11.2	144.4
Fuel Combustion in stationary sources	24.4	7.2	1.8	0.9	10.0	44.3
Industrial processes	7.5	14.4	12.0	5.5	0.2	39.6
Solid waste disposal	0.2	1.4	7.9	2.0	0.4	11.9
Miscellaneous	0.2	11.4	18.2	9.2	2.0	41.0
TOTAL	33.4	35.2	151.4	37.4	23.8	281.2

Note: Sulfur oxides are expressed as sulfur dioxide and nitrogen oxides as nitrogen dioxide in this table.
Source: Environmental Protection Agency.

In addition to the above pollutants, many others plague the environment. For example, asbestos and the aerosols. Asbestos, a fibrous mineral substance, is used in construction, insulation, and brake linings for automobiles. It is important because it retards fire. It has been used for decades, but recently it has been found that asbestos dust getting into the lungs causes a form of cancer known as mesothelioma. We are all exposed to it, since it is found in almost all buildings. In addition, every time a car brakes on the street, asbestos is emitted into the air. The difficulty with this material is that we do not know of a "safe" level of exposure, because it takes years or even decades for trouble to develop. This most certainly influenced the EPA's decision to classify asbestos as a hazardous air pollutant in the 1970 Clean Air Amendment.

Aerosols in recent years have been widely discussed because of their effects on the ozone layer. What does this mean? To simplify, the aerosol cans that contain deodorants, hair sprays, and air fresheners all contain freon compounds. As we increase the use of this material on earth, it rises in the air and collides with intense solar radiation twenty miles above the earth. This in turn releases chlorine atoms (present in freon compounds). The chlorine atoms fall back down to the ozone layer, about fifteen miles above the earth. Chlorine atoms (Cl_2) join the ozone molecule (O_3), giving way to free oxygen (O_2) and forming chlorine oxide molecules (ClO). A free oxygen atom may now hit chlorine oxide molecules, forming molecular oxygen and once again freeing chlorine atoms for forming the process again.

$$ClO + O \text{------} O_2 + Cl_2$$

In this way, we are changing the ozone layer (O_3), which protects the earth from the deadly ultraviolet rays, to a molecular oxygen layer (O_2), which permits ultraviolet to pass through. If this continues, living things on the earth may burn up. Currently, many companies are substituting CO_2 for freon compounds in aerosol cans; to date there is no evidence of environmental problems from these aerosols. In this way, industry has met the new challenges of providing a "safe" product for both the environment and man.

Tobacco smoke in enclosed spaces, just like any other form of undiluted smoke, must be recognized as an air pollutant. The nonsmoker, who often inhales a tobacco smoke–polluted atmosphere, is particularly affected by this form of pollution. Recent evidence suggests that the nonsmoker living in an environment with smokers incurs the same health hazards as the smoker. Of course, the exposure of the nonsmoker depends upon the amount of smoke produced, the depth of inhalation by the smoker, the ventilation available, and the closeness of the burning tobacco. The length of time a person is exposed to tobacco smoke determines how much of the pollutant is absorbed into the body.

Tobacco smoking releases a large variety of toxic substances into the air. Among these pollutants are CO, nicotine, hydrogen cyanide (HCN), nitrogen dioxide (NO_2), and tars. The pollutants considered the most hazardous to health are CO, nicotine, and tars. We have already seen the effects of CO, and further research shows that there is no level of CO in ambient air that is without some effect. It is estimated that cigarette smoke contains about 42,000 ppm of CO; this is, of course, diluted in the smoker's lungs and also diluted in the air surrounding the smoker. However, concentrations of 20 to 80 ppm of CO have been found in enclosed spaces, such as a smoke-filled room. This surely can have an effect on the respiratory and circulatory systems.

Nicotine has been implicated as a cause of heart disease among smokers because of its diverse effects on the circulatory system. It is aggravated by the presence of CO. Tars, a particulate mixture, include many cancer-inducing substances. The most abundant single carcinogen is benzopyrene, found in cigarette smoke. It has produced tumors on the skin of mice simply from the sidestream smoke of a cigarette burning in an ashtray.

Evidence that tobacco smoke pollution is a hazard to nonsmokers is mounting. Nicotine appears in the blood and urine of nonsmokers exposed to indoor tobacco smoke. Children of families where smoking occurs have a significantly higher incidence of respiratory ailments than children of nonsmoking families. In all, we must recognize the need for, and the development of, programs that foster a safer environment and promote life-styles conducive to health.

What is being done about air pollution control? Obviously not enough. We are only now realizing the extent of air pollution and its seriousness. In the past, the Clean Air Act (1963) and the Air Quality Act (1967) did very little to stop serious pollution. With the passage of the Clean Air Amendment (1970), there are stronger legal tools for air pollution control and an even larger mandate for citizen participation. In this amendment, Congress finds that the prevention and control of air pollution at its source is the primary responsibility of state and local governments. However, if the states fail to meet their control responsibility, the federal government, acting through the EPA, has the responsibility and authority to enforce pollution

A waste incinerator plant.

control. Thus, we now have a monitoring system of the air, standards of performance for pollutants, auto emission controls, a classification of hazardous air pollutants including asbestos and mercury, and federal enforcement in order to protect man and his environment from such pollution. It must be remembered, however, that any law is only as good as the people enforcing it. Only with an understanding of the causes and symptoms can we hope for a brighter tomorrow in which to breathe.

VI

Impacts of Water Pollution

POLLUTION

TOM LEHRER

If you visit American city,
You will find it very pretty.
Just two things of which you must be-
    ware:
Don't drink the water and don't breathe
    the air.
Pollution, pollution,
They got smog and sewage and mud,
Turn on your tap and get hot and cold
    running crud.

See the halibuts and the sturgeons
Being wiped out by detergents.

Fish got to swim and birds got to fly
But they don't last long if they try.

Pollution, pollution,
You can use the latest toothpaste,
And then rinse your mouth with indus-
trial waste.

Just go out for a breath of air,
And you'll be ready for Medicare.
The city streets are really quite a thrill,
If the hoods don't get you, the monoxide
will.

Pollution, pollution,
Wear a gas mask and a veil.
Then you can breathe, long as you don't
inhale.

Lots of things there that you can drink,
But stay away from the kitchen sink.
Throw out your breakfast garbage, and
I've got a hunch
That the folks downstream will drink it
for lunch.

So go to the city, see the crazy people
there.
Like lambs to the slaughter
They're drinking the water
And breathing the air.

The hydrologic cycle (Fig. 8) enables the U.S. to
have about 30 inches per year of rain, or approximately

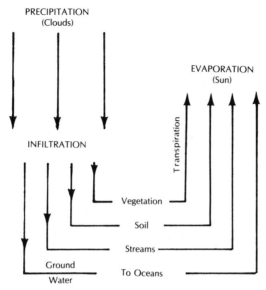

FIG. 8

4,300 billion gallons of water per day. Of this amount, about 3,000 billion gallons per day are lost to the atmosphere through evaporation. The runoff, then, is approximately 1,300 billion gallons per day. This is the water available for man to use as he sees fit. We must remember, however, that this water is available only in certain parts of the country, especially the Southeast and the Pacific Northwest. Many scientists assert that before the year 2000 a permanent water shortage affecting our standard of living will occur. Water is used in almost everything we eat and do. We use it for swimming, bathing, preparing food, agriculture, and meat production. Yet we still take this precious resource for granted. Just think of the waste of water and, more important, the polluting of the waterways. All over the country, freshwater lakes and streams are being converted into sewers because of man's lack of understanding of the water system. A few hundred years ago, if a man swimming in a stream in the countryside urinated in the water, it would have had very little, if any, effect on the water. It would be diluted out. At present, if people were to do the same thing we would have a sewer to swim in, since the number of people swimming in the same body of water has increased so tremendously. No longer would dilution be able to take care of it.

Our lives and the way we live are dependent on a steady flow of clean water. According to the U.S. Department of Commerce, each person in the U.S. uses between 50 and 200 gallons of water each day. When the water used by industry and agriculture is added to the total, the amount of water needed by each of us rises to about 2,000 gallons per day! Let's take a closer look at some of the ways water is utilized. Water is used on the farms to grow the food we eat. For example, 5 gallons of water are needed to produce 1 gallon of milk; 2,000 tons of water are needed to produce 20 tons of fresh vegetables. Water is also used in power plants to produce electricity. We use about 80 gallons of water to generate about 1 kilowatt hour of electricity (1,000 watts of electricity per hour or ten 100-watt bulbs turned on together for one hour). In addition, water is used in manufacturing. To produce

1 ton of steel needed in construction takes about 65,000 gallons of water.

So, as you can see, the way we use the water given to us through the hydrologic cycle is important. We therefore must learn to keep the waters clean and available for use by man and other living things if we are to survive on this planet. Too often, however, we pollute the water and thus make it not readily usable. What are some types of water pollution that we can see and analyze?

There are four basic types of water pollution. The first is *chemical organic pollution*. This consists of oils, dyes, detergents, and acids that are all "misplaced" and put into our environment. The second type is the *inorganics,* which include chemical compounds such as nitrates, phosphates, and sulfates. The third type is *physical,* including floating matter, suspended matter, or thermal matter; in other words, such things are leaves, wood, foam, or heat. The fourth type of water pollutant is *biological,* which includes living matter such as algae, weeds, bacteria, and worms.

The main problems with pollutants in water is that they may cause a reduction in the complexity of the web of life in the waterway. In other words, they might affect the *number of different species* available in the body of water. This is called *species diversity.* It is a most important concept of ecology. If a body of water becomes *less* diverse or the number of species becomes fewer, the body of water becomes too simple and usually unhealthy. Keeping the body of water healthy requires many species that can live together to form a complex web. For example, a lake may contain 200 species of organisms (trout, bass, Elodea plants, and others). Each of these 200 species has 5 members, making a total of 1,000 individuals. This is considered healthier than a lake that contains only 2 different species with 500 members in each, making the same total of 1,000 individuals. Why? Obviously, if the lake contained 1,000 individuals representing only two species and the environment should drastically change, then the 2 species might be wiped out because they might not be able to adapt to the changes. However, in the lake with 200 species there is a better chance that some of the species can adapt to the changes in the environment, and the lake will remain functional.

There are several ways to classify water pollutants in more specific terms. The following is the system used by the Environmental Protection Agency.

Oxygen-demanding wastes. These come from living matter, food-processing plants, and various manufacturing processes. When these wastes are broken down by bacteria, oxygen is removed from the water. If the oxygen level drops too low, some fish will die. Thus, we must be able to measure the oxygen required to sustain life in the water. This is called the BOD (Biochemical Oxygen Demand). In 1971, the EPA reported that 73.7 million fish were killed by water pollution in the U.S. The main source of pollution was from sewage that had low dissolved oxygen in the water, which caused the deaths.

The amount of DO (dissolved oxygen) in water is important in determining the quality of water. Fish, for example, require a minimum amount of DO depending on the level of activity, water temperature, and stage of development. The more active the fish, the more oxygen it uses. The warmer the temperature of the water, the less oxygen is available to the fish. All are important in determining whether or not a fish can survive in a body of water.

Disease-causing agents. These come from human and animal waste, which usually enters the water with sewage. Disease-causing microbes can come into contact with humans through drinking water.

A simple procedure has been developed to test water for microbes. Millipore Technology has advanced techniques that enable environmentalists to analyze water samples quickly and efficiently.

POLLUTION DETECTION BY MEMBRANE FILTRATION

Extension of the Capabilities of Teachers and Students in Laboratory and Field Experiences

by BERNARD I. SOHN

Every so often a new technology emerges that presents a refreshingly simple and new approach to solving certain scientific problems. Membrane filtration is such a technology.

The precise separations made possible by membrane filters enable workers in science and industry to perform tasks that were once thought to be impossible or impractical. Likewise, precise separations with membrane filters permit students of environmental science more easily and positively to isolate microorganisms from air, water, soil, and foods. Membrane filters also simplify analysis of discrete particles when determining such pollutants as pollen levels, particulate loading of cigarette smoke, degree of lead content in automobile exhaust, and the amount of radioactive particles in air. Even the elementary student can undertake all the above investigations using membrane filters.

The technology of membrane microfiltration was an outgrowth of World War II. The membrane filter was developed to provide a quick, reliable means of detecting bacteriological contamination in air, food, or water supplies in the event of an attack using biological weapons. Air and water pollution analyses continue to be important applications for membrane filters. The microporous plastic membranes, which gave rise to the technology, have many unusual properties—properties which lend themselves remarkably well to the dramatization of environmental science in the school laboratory as well as on field trips.

Perhaps a brief look at how this unique filtering material is made might be in order:

Two kinds of plastic are mixed and formed into a polymeric sheet containing billions of uniform pores or holes. The molecules of one of the two plastics are stable. Molecules of the other kind are relatively volatile, but they can be stabilized as desired. A uniform pore structure is produced as the volatile molecules of one are forced to volatilize under controlled conditions, leaving gas-tracks (capillary pores) in their path as they escape through the polymerized sheet formed by the molecules of the other. Membrane filters can be produced in a range of sizes down to a diameter of a 0.025 micrometers. Pores of this minimum diameter are beyond the resolving power of optical microscopes and will hold back particles as small as a polio virus. Although the membrane is actually clear, it appears to be white because the billions of perforations scatter light. However, when the pores are filled with a fluid such as microscope immersion oil, the membrane becomes transparent. Very simply, the air in the pores is replaced with a fluid of matching refractive index (1.5). Twenty percent of the membrane is solid; eighty percent is empty space. By contrast, ordinary window screening is less than 55 percent porous.

The key to the *technology* lies in the unique way that a membrane filter "screens out" particles. Ordinary filter paper, which is essentially a nonuniform fibrous material, will trap particles throughout the entire depth of the material. Some particles, particularly individual bacterial cells, will find their way through the maze of fibrous material and emerge in the so-called "filtered fluid."

By contrast, a membrane filter which is a thin plastic disc (about the thickness of fine writing paper) works by "screening out" particles on its surface. It traps and holds all particles including bacterial cells in a single plane of focus allowing high magnification and identification of the entrapped particles or microorganisms.

Given this unique material, let us explore some of the ways it can be incorporated into an environmental science program. The basic equipment and simple techniques described can be applied across a cross section of investigations at many grade levels depending on the degree of analytical sophistication employed.

Equipment and Techniques

The equipment needed for virtually every membrane filtration investigation is a vacuum filter holder, a vacuum source, a pair of nonserrated forceps, and a membrane filter disc of the appropriate pore size and diameter. The most versatile type of filter holder for environmental investigations is a Sterifil^R filter holder.* It is a modified Büchner funnel of sorts, is made of autoclavable plastic (Lexan polycarbonate), and will withstand rugged use (or abuse) by students.

A typical series of manipulations might be as follows:
1. Unscrew the funnel portion of the Sterifil and, using smooth-tipped forceps, place a membrane filter in position.
2. Replace the funnel.
3. Attach a vacuum source, such as a hand vacuum assembly.
4. Add a volume of water or other liquid to be filtered to the Sterifil funnel.
5. Apply vacuum and draw the liquid through the membrane.
6. Release the vacuum, carefully remove the filter, holding it with smooth-tipped forceps, and proceed to analyze the filter in accordance with the objective of the specific investigation.

Having established how the Millipore membrane filter works, let us now explore a few of the many applications of this technology in an environmental science program.

1. Experiments with Water

A local pond or river can furnish abundant opportunities for ecological investigations, including those for protozoans, phytoplankton, and coliform bacteria.

a. Life in a drop of pond water

A classical experiment in beginning biology is that of microscopically observing unicellular life in a drop of pond water. Sometimes, however, the population density of the protozoans is quite low, leading a student to get somewhat frustrated while trying to find a "little critter" to focus in on. However, applying the membrane filter technique, a sample of pond water can be "concentrated"

---

* Available from Millipore Corporation, Bedford, Massachusetts, USA, with subsidiaries in most major countries.

by drawing a portion of it through the filter; releasing the vacuum will stop the filtration. By this method, a volume of 100 ml pond water can be concentrated tenfold with a corresponding increase in the population density. A drop of this concentrated sample is now "teeming with unicellular life" and provides the student with some very "exciting" viewing.

b. Phytoplankton Analysis

The "number" and "types" of algae in a pond water sample provide the student investigator with a profile of the water quality. In polluted waters, certain species such as blue-green algae and flagellates become more prevalent while the number of green algae and diatoms tend to diminish. By filtering a sample of pond water the student can collect all algae on to a single microscopic plane on the membrane surface. The filter is then placed on immersion oil and rendered transparent. By this method, even the smallest morphological detail becomes accessible to the microscopic eye, and since the filter is imprinted with a network of grid lines, statistical enumeration of the algae density becomes a routine exercise.

c. Coliform Analysis

The single most important biological indicator of bacterial pollution in water is a simple very common class of organisms known as coliform bacteria.

As one microbiologist put it, looking for bacteria in water is like looking for pigs in the dark: it's easier to find the pig that squeals. This, he explains, is the reason sanitarians and other water-pollution analysts first look for the harmless coliform bacteria in water rather than the real troublemakers: disease-producing organisms that pose a real threat to human health.

The attribute that makes coliform bacteria so easy to detect is their special ability to break down a complex "sugar" called lactose to form several simpler substances, one of which will combine with a fuchsin stain (an ingredient in the culture medium) to form an iridescent green coating over the coliform colony. These colorful "sheen" colonies are easy to distinguish from their less colorful noncoliform counterparts (the pigs that do not squeal).

There are other reasons, though, for the choice of coliform as the official criterion of the sanitary quality of water. First of all, coliform bacteria usually orginate in the intestines of warm-blooded animals, including man. Therefore, their presence in water in *unusual* number is cause for concern, since human wastes are the most likely source of organisms pathogenic to man, such as those that cause typhoid fever, dysentery, or cholera. Coliform bacteria are also hardier than most pathogenic organisms; thus it is unlikely that the pathogenic species are still surviving if the coliform levels are low. Coliform grow readily at ambient room temperatures, so samples can be cultured easily without a laboratory incubator. Ideally, however, coliform grow best at 35°C.

The simplest and most widely used test for coliform bacteria consists of filtering a water sample through a sterile bacterial-retentive membrane filter. The microscopically small filter pores let the water through leaving the organisms trapped on the filter surface. At this point, the microorganisms are invisible to the naked eye; but when the filter is placed on a paper pad soaked with nutrient Endo medium, the nutrients wick up (capillary action) through the filter pores to keep the microbes fed and multiplying. The nutrient Endo medium contains lactose plus basic fuchsin. The selectivity of the test is achieved in the color reaction that takes place when coliform organisms break down lactose to form an acid aldehyde. The latter metabolic end product reacts with the basic fuchsin to give the coliform colonies a characteristic green sheen. Completed cultures can be dried, deactivated, and preserved as permanent records of the experiment. It is important to stress that in the above experiment it is highly unlikely that students can culture any pathogens, but rather only the *indicator organisms*. Nevertheless, careful laboratory techniques and handling practices should be stressed.

Plant nutrients. These are compounds of nitrogen and phosphorus that come from sewage, industrial waste, and land drainage. They feed many water plants in excess of daily requirements. Agricultural runoff contributes to large amounts of these nutrients, usually in the form of artificial fertilizers that cause rapid and excessive amounts of algae to grow. This process is known as *eutrophication.* One can see its effects on fresh water by the rapid filling in of the water by the vegetation, eventually displacing the lake and forming a possible marsh.

Toxic substances. These are found in the form of synthetic organic compounds. Some examples include detergents, cleaning agents, and "bug" killers. Many of these substances are toxic to aquatic life and to human health. Inorganic substances such as salts and acids cause problems in changing the pH of the water.

Persistent substances. These are substances that are not normally biodegradable, or broken down easily in nature. They get into the water from runoff, agricultural practices, and industrial operations. If they remain in the environment without being decomposed, they can accumulate and cause havoc to life cycles.

Sediments. These include particles of soil, sand, and minerals that are washed off the land into the water. They settle to the bottom of waterways and require dredging to clear them off. They also reduce fish and shellfish populations by covering their food sources. If they are organic, they may decompose and use additional BOD (biochemical oxygen demand).

Heat. This comes from electric power plants; it may raise the temperature of the receiving water by as much as 20° F and cause changes in the life cycles in the water. Warm water holds less oxygen and absorbs less from the air. Thus, the amount of DO available in warming waters is less than in cooler ones.

Radioactive substances. These substances enter the water from the mining and processing of radioactive ores. They also result from fallout if nuclear weapons are tested in the atmosphere. Radioactivity concentrates in the food chains and can cause health problems.

Population and industrial expansion during the past fifty years has vastly increased the volume of potential water pollutants, so that nature's self-cleansing capabilities are overwhelmed. When this is so, the problem is man-made and demands man-made solutions.

VII

Impacts of Noise Pollution

The impact of noise on the environment is becoming more and more recognized, although until recently it was not even considered a form of pollution. *Noise pollution* is one of the unwanted side effects of man and industry. It has been with us for centuries, but currently it is becoming more prominent as the sounds become greater with the population increase and the expansion of industries.

Noise in the past was not considered a pollutant because it differs from the other pollutants we have discussed. First, it does not create any "storage problem"; that is, it is dissipated immediately, whereas other pollutants from the air and water can accumulate. Second, noise is by and large a "people's problem." It does not appear to be involved in the interactions of other living things and their life cycles.

Noise can be considered as unwanted sound, even though the term "unwanted" is difficult to define. What might be unwanted to one person may be wanted by another. We must, therefore, think of noise in terms of its psychosocial aspects. That is what makes this pollutant difficult to study, since one person might enjoy one type of sound, such as a rock concert, while another considers it just noise. Why noise? It is annoying to that person. Look at it another way. You might consider the sound of an opera as noise, while a friend might consider it pleasurable. So you can see that it is difficult to define what noise really is. There is, however, another way of looking at noise. Noise can be considered in terms of health and physiology. Recently, noise has been redefined as any audible sound that might be detrimental to health. With this new definition, we can see some of the effects sounds have on human well-being. It has been found that some environmental noise levels have the effect of producing temporary or permanent hearing loss; physical and mental disturbances; and interference with voice communication, job performance, and sleep.

There is much current speculation as to whether various noise levels can cause physical and possibly mental disorders. Some reports indicate that loud noise causes the constriction of small capillary blood vessels in the body, reducing the flow of blood, and in some people raises blood pressure and alters the heartbeat rate.

In order to better understand these effects we should understand how sound is produced and transmitted. Whenever something is caused to vibrate and the vibrations are picked up by the ears, we say we hear a *sound.* The outer ear picks up the vibrations from the surrounding environment and transmits them to the ear canal, to the eardrum, and on into the middle ear.

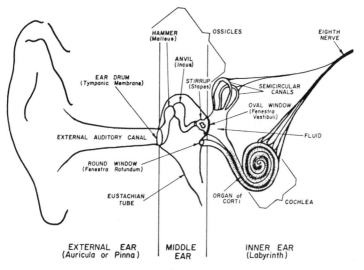

FIG. 9

PHOTO BY LOUIS SALZBURG

Instruments at the Bureau of Noise Abatement in New York City.

81

Chart of Noise Levels

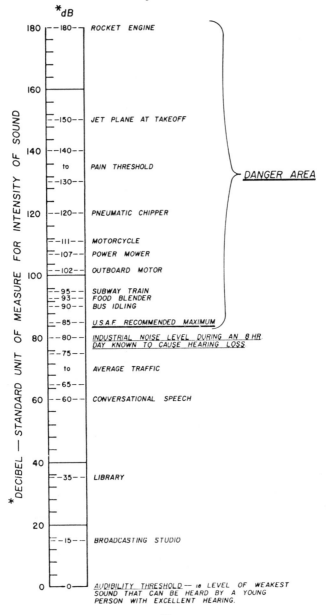

*dB

180	— 180 —	ROCKET ENGINE
160		
	— 150 —	JET PLANE AT TAKEOFF
140	— 140 —	
	to	PAIN THRESHOLD
	— 130 —	
120	— 120 —	PNEUMATIC CHIPPER
	— 111 —	MOTORCYCLE
	— 107 —	POWER MOWER
	— 102 —	OUTBOARD MOTOR
100		
	— 95 —	SUBWAY TRAIN
	— 93 —	FOOD BLENDER
	— 90 —	BUS IDLING
	— 85 —	U.S.A.F. RECOMMENDED MAXIMUM
80	— 80 —	INDUSTRIAL NOISE LEVEL DURING AN 8 HR. DAY KNOWN TO CAUSE HEARING LOSS.
	— 75 —	
	to	AVERAGE TRAFFIC
	— 65 —	
60	— 60 —	CONVERSATIONAL SPEECH
40		
	— 35 —	LIBRARY
20		
	— 15 —	BROADCASTING STUDIO
0	— 0 —	AUDIBILITY THRESHOLD — 10 LEVEL OF WEAKEST SOUND THAT CAN BE HEARD BY A YOUNG PERSON WITH EXCELLENT HEARING.

DANGER AREA

*DECIBEL — STANDARD UNIT OF MEASURE FOR INTENSITY OF SOUND

These vibrations are transmitted from the middle ear, where the eardrum causes three small bones to vibrate, to the fluid in the inner ear, where tiny hair cells are located. The vibrations of the hair cells and the fluid cause an electric signal to be transmitted to the brain through the auditory nerve that ends in the inner ear.

How can we measure the sound received by the ear? The *decibel* (dB) is the unit of measuring the intensity of a sound. It should be remembered that loudness is a subjective reaction of the listener to the intensity of a sound as it is received, not to the power of the source. This unit is a logarithmic scale in which the rating of zero represents the threshold of human hearing. In addition to intensity, there is the tone level or pitch of sound. Pitch is determined by how many times per second the sound wave vibrates and is measured in hertz (Hz). The high frequency noises that we can hear seem louder to us and are more disturbing than the low frequency noises.

The most specific effects of noise are on hearing. A loud blast such as an explosion or a jet taking off at an airport deafens everyone for a moment. Permanent hearing loss comes only from repeated or continuous exposure. It is agreed that an 8-hour daily exposure of 80 to 85 dBs leads to severe hearing damage. Until recently this was thought to be the only damaging effect of noise on the ears. Currently, however, doctors agree that noise that surrounds us, especially in the urban environment, can cause deafness. Such noises include traffic, planes, car horns, and air conditioners. In addition, these noises are believed to have effects on stress levels that may affect mental well-being.

Noise is a true environmental pollutant. Like air and water pollution, it damages health and lowers the quality of life. In Europe the malignant effects of noise

Noise pollution: The roar of a row of vehicles causes the passerby to try to block out the sound.

on the total environment are well recognized. In Sweden, for example, the State Power Board has a one-year course in pollution control to train engineers to deal with water, air, and noise problems. In the U.S. we are barely beginning to realize what a growing menace noise pollution is and that it must be controlled.

Noise pollution: Sidewalks must be opened up for repair of utilities, but the concomitant noise can be painful to nearby residents.

Noise pollution: Life near a busy airport can be hazardous to one's hearing.

Noise pollution: On a hot summer day, the whirr of countless air conditioners can make the street even noisier than usual.

VIII

The Impacts of Biocides

When man places a chemical into the ecosystem to rid himself of another living thing, we say he is using a *biocide*. In our environment we encounter various types of biocides. Those against races of man are called genocides, those against pests are called pesticides. For centuries man has been plagued by pests that either threaten his crops and food sources or cause health problems. Some such pesticides that have been developed are herbicides (against herbaceous plants), arboricides (against woody plants), insecticides (against insects), and arachnicides (against spiders).

It is, however, the insect pests that have bothered man more than any other living thing and have threatened his existence over the centuries. To deal with these pests, man has developed biocides. In doing so, however, he has created other ecological imbalances and impacts. Insects are probably the most successful animal (invertebrate) on the earth and have in many instances out-smarted man.

First, insects are *adaptable*. They are able to live in wide ranges of environmental conditions—such as extremes in climate, water or land—and they can eat almost anything, including human hair and nails. Sec-

87

ond, their *external skeleton* is resistant to many chemicals and also prevents them from drying out. Third, their *ability to fly* allows for a wider distribution of their numbers and a greater choice of food and environments. Fourth, they are able *to change their form* (metamorphosis) as environmental conditions are altered and as a result can survive in different habitats. They also eat different foods in various stages of their life cycle. For example, the first stage in the life cycle of a butterfly is the laying of the eggs. During the second stage the eggs hatch into larvae or caterpillars, which chew leaves and stems and are destructive to vegetation. The third, or resting, stage is that of the pupa, in which a cocoon is formed. The last stage, complete metamorphosis, is the change from pupa to adult—the reproductive stage. Here new eggs are produced, and the cycle starts all over again. Lastly, insects are so successful because of their specialized reproductive system, which enables sperm to be stored for long periods of time. This allows for the best environmental conditions to be present before eggs are laid. In addition, egg production is so large that one female insect may lay hundreds of thousands of eggs in one season. This of course is important for the species, since it offers a greater chance for variability and mutation.

Thus it is clear that insects are quite successful. Not only do they reproduce in huge numbers, but also the variety that prevails enables them to adapt to all kinds of changes within their environment. Unfortunately, man has always thought of insects as pests, even though only about 900 species of insects out of a million or so are actually pests to man. In order to deal with these "pests," man has spent time and money developing biocides to rid the environment of these creatures. Obviously, one cannot completely get rid of all insects because of their successful adaptations and mode of living. But man still tries to deal effectively with the problem. In many instances he does not take into account the impact he is making on the environment. Let's look at some of the insecticides man has developed to rid himself of insect pests.

Five types of insecticides are generally recognized:

Naturally occurring products. These are chemicals usually obtained from plants. In the past we have used chemicals from the dried flowers of daisies and chrysanthemums. In addition, nicotine and garlic also have proved successful.

Inorganic Chemicals. These are chemical preparations containing metals such as lead, mercury, and zinc. These generally are not used because of the residual impact they have on life in the environment.

Organochlorinates. These are also known as chlorinated hydrocarbons. The best known and most widely used organochlorinate is DDT (DichloroDiphenylTrichloroethane). It has been used to combat diseases such as malaria, yellow fever, and plague and to save crops. Other organochlorinates include lindane, dieldrin, and mirex, which are commercially available in garden shops. As a group, the organochlorinates are persistent in the ecosystem, have a doubling effect in the food chain, and accumulate and circulate within the ecosystem (see Fig. 11).

At present, DDT is outlawed in most of the states

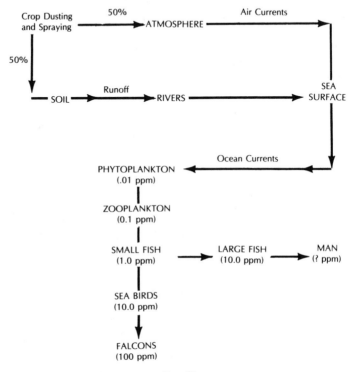

FIG. 11

because of its nature and effects on the environment. However, the rest of the world is still using this insecticide, and it still is circulated throughout the world, since our planet is a closed system.

Organophosphates. These chemicals include malathion, parathion, and diazinon. They are all broad-spectrum chemicals (kill almost all insects that come in contact with them), but they are not persistent in the ecosystem because they are biodegradable in a short period of time. However, they are dangerous to use with the chlorinated hydrocarbons because of known poisoning effects.

Petroleum Oil Fractions. These are oils that are usually used alone or with one of the other types of insecticides described. They are most useful in controlling mosquito larvae. The oil is spread as a thin film on the surface of water where mosquitos breed; as the larvae come to the surface to feed and breathe, they contact the oil film which penetrates their breathing tube and suffocates them.

Over the many decades man has successfully used one or several of these types of insecticides to control pests. In many instances they have helped to save lives, such as control of malaria and yellow fever in the tropics. They have saved many fruit orchards and crops from the hungry insects that compete for man's food.

Because of the insects' unique features and adaptability, however, man is finding it more and more difficult to control them. We have lost sight of the fact that as we use chemical insecticides, the insect pests build up tolerance levels and resistance to the chemicals being used. This can take place in a rather short time because of the rapidity with which insects reproduce. Man, therefore, begins to fall into the "insecticide trap," in which we increase the dosage of chemicals used to control the increase in insect pests over the years.

As new strains of insect offspring develop, many are tolerant to the dosage of the chemical and thus have a built-in resistance to it. To counteract this, we must over the next year either increase the dosage to control this "new breed" or find a new chemical to which the insect pest does not have a resistance. And so it goes, on and on until eventually there is no chemical to control the pests or until we find a chemical that may also kill man. Currently, for example, fruit orchards are being sprayed eight to nine times a year, compared to ten years ago when they were sprayed only three to four times.

With this increase in use and dosage, what are the effects on man and the environment? Experiments have been performed to show that by using broad-spectrum insecticides we kill off the "good" insects that can naturally control the "pests" through the food chain. These chemicals do not discriminate between good and bad; they simply kill all the insects. Thus, we are altering the food chains within the ecosystem and making impacts on all living things. Furthermore, even though there is no scientific evidence to prove that insecticides have directly killed man, we do find that many are cancer-producing agents. We also know that some types of insecticides accumulate within the food chains, and perhaps as the quantity of chemical accumulates it might cause other health problems.

Because some insecticides are persistent in the environment, we also have difficulty in getting rid of them. They remain in the soil and water for many years, and we don't know what effects they have on us or on the rest of the environment. Thus, insecticides surely make an impact on the environment in many ways. What is the answer?

In attempting to control pests, we must identify the problem that the pest causes and analyze all the available information on the pest and the damage it causes in order to determine what action to take. In many instances, it might be best to let nature take care of it. We must remember, however, that we are to blame in part for the present impact that insecticides are having on the environment. For example, when we go to the store to buy an apple we select only those apples that have no blemish or spot. We want "perfect" apples. This requires a great deal of time and expense in using chemicals to assure the apple grower that he can have almost all of his crop free from such marks. This will bring him a greater profit and will please the consumer. However, each year he must use more and more insecticide at greater expense to the consumer in many ways to provide a blemish-free apple. We are also to blame in our attitude, and perhaps we should try to change it and accept an apple with a spot or worm hole in it.

Furthermore, we do not follow directions when using chemical insecticides. Too often when using an insecticide the directions tell us to use 1 teaspoon of chemical to 1 gallon of water. What do we do? We use 2 teaspoons to 1 gallon. We want to make sure the insect will be more dead! This attitude must change if we are going to survive in harmony with our environment.

Let's now look at some alternate methods we may use to help in our fight against insect pests.

Biological control. Use of a prey-predator relationship, setting up a natural ecological balance with food chains. This involves knowing the pest's behavior and using it to control its numbers and patterns of living.

Ecological selectivity. Use of chemical pesticides that are narrow spectrum, not broad. These are specific insecticides for a specific insect pest. For example, to dispose of gypsy moths develop a chemical that will only affect the gypsy moth and not harm other insects.

Research. We need more research on the effects insects have on various environments. Perhaps we need to learn how to use insects to benefit mankind, rather than looking at them only as pests. We also need to develop controls that will have the least impact on the environment and yet affect the pests. This will require dedicated scientists to develop and test new ideas of pest control.

Education. Through understanding of the total environment and the impacts of chemical controls, we should be able to make better decisions on how to use such products. Too often we only look at the immediate effects, and not the long-term effects. It is time to analyze such materials in greater depth to see how they affect man and his environment.

In addition to biocides, we should also be aware of other important chemicals that are included in more and more foods today. These foods have been grown or processed in some way that includes contaminants or additives. *Contaminants* are chemical compounds that are retained in food as either biocide residues on plants or antibiotics or hormones in meat. *Additives* are substances that are specifically added to foods during their preparation or processing to assure longer shelf life, flavor, attractiveness, or ease of preparation.

Biocides have been discussed above; let's study some other contaminants:

Antibiotics. It has been found that application of antibiotics to livestock can act as a growth inducer and also prevent diseases. Fed the growth inducer, livestock can be marketed sooner at a greater weight and profit, although perhaps contaminated with the antibiotics. The danger in using antibiotics as a growth inducer is twofold. First, people who are sensitive to antibiotics may be exposed to such chemicals without their knowledge; and if people are constantly and unknowingly exposed to antibiotics, resistant forms of microorganisms may develop, making the drugs useless in a time of real need. Second, recent evidence indicates that drug resistance is a genetic factor that can be transferred from one organism to another. Thus, if nonharmful bacteria found both in man and animals could acquire and pass on the resistance factor to any of several harmful bacteria normally affected by antibiotic drugs, then man would not have any method of controlling these harmful microbes.

Hormones. Through development of hormonal contraceptives, the compound Stibestrol was synthesized, which causes weight gain in animals such as chickens. For example, a small amount of this hormone placed under the skin of the fowl will cause it to become plump and fat. Recent experimental evidence has shown that eating animals that contain this hormone can cause an imbalance of sex hormones in human bodies and possibly cause cancer. Again, we can see that as we try to control one factor in our environment, we unbalance others. We need to rethink our ways in terms of how one thing is connected to everything else.

The food industry has used additives to help solve a problem with food, namely, the short storage life. This has been an age-old problem during which salt, sugar, and spices were used as preservatives. Today, however, nitrates and nitrites are commonly used to preserve meat products; benzoic acid is used to preserve liquids that might ferment; and sulfur dioxide is used to preserve dried fruits. All such practices lead to addition of artificial colors and flavors. We have developed an industry that is adding so many chemicals to foods to enhance their flavor, provide a richer color, and help in preservation that soon we will be simply eating just the chemicals! Research in this field has shown that many of the additives, if used in quantity, can cause cancer and other health problems. Let's study a few:

Artificial Sweeteners: Saccharin, discovered in 1879, has been used for years as a sugar substitute, especially by diabetics who need to control sugar intake. Saccharin is not metabolized by the body and is considered safe within recommended doses. But, again, one needs to remember that it is a nonfood item. Cyclamates, on the other hand, are apparently metabolized by some people and have been shown to be a breakdown product. For example, when cyclamates were applied to rat cells, they caused chromosomes to break up. This in turn has genetic effects. Cyclamates have also been cited as a possible cause of birth defects in birds.

Food Dyes. Of all the food additives, dyes made from coal tars are considered the most dangerous to health. It has been found that dyes to make oranges orange, hot dogs red, and cherries scarlet are cancer-causing. Is it really important to have our cherries more red than red? If so, look what has to be done—maraschino cherries are preserved with sodium benzoate, made firm with calcium hydroxide, bleached with sulfur dioxide, artificially flavored, and then given their bright red color with a coal tar dye. All of this just for a beautiful cherry! A rather high price to pay.

It is time we found out more about such additives and protected ourselves from materials that might prove hazardous to health.

IX

We Can Save the Environment

In previous chapters we have discussed various problems facing man and his environment. In this chapter we shall try our hand at solving some of these environmental problems, but first let's review some "basic laws" of the environment that help to explain the many interactions between man and the biotechnosphere.

Law #1. *Everything is connected to everything else*

This law refers mainly to the various cycles that interact within the environment and the recycling of materials throughout the ecosystems. If we upset one part of the cycle or change it, the rest of the cycle is thrown out of balance. For example, a great drought that causes hardship to farmers also affects the people living in eastern cities.

Law #2. *Everything must go somewhere*

This law refers to the earth as a closed system. People say, "Throw it away," but there is no "away." What happens on one part of our planet affects another part.

For example, when you throw a mercury battery into a garbage can for incineration, the mercury is released and eventually ends up in waterways, where it changes into a poisonous compound. Fish can store this compound, and it eventually is given back to us when we eat the fish.

Law #3. *Nature knows best*

Nature is able to decompose, recycle, and restructure all those factors that are needed for survival in a healthy environment. Any major man-made change in the natural system is likely to be bad for the system. Many new substances have been manufactured and sold without regard for the effects they may have on nature and man. Some examples are plastics, DDT, artificial red coloring, spray cans, and detergents.

Law #4. *There is no such thing as a free lunch*

This "law" simply means that every gain is won at some cost. Every time we take something out of the environment, or change it, we must pay a price. For example, for us to have electricity, we must use fossil fuels, a natural resource that is limited.

Let us now use these laws and the other concepts discussed in this book as we examine a model of the earth's closed system.

Next to each of the questions posed, indicate whether

you (1) strongly agree, (2) agree, (3) disagree, or (4) strongly disagree, and give the reasons why you chose that decision.

In the world today, approximately two-thirds of the people are desperately poor, and only one-third are comparatively rich. The people in the poor countries have an average per capita Gross National Product of about $200 per year; in the rich countries, of about $3,000 per year (in the U.S. it is about $5,000 per year). Let us say that the rich countries fill a lifeboat with all the rich people, and the poor countries fill a lifeboat with all the poor people. Since there are many more poor people, their lifeboat spills over and many must swim for their lives. While in the water, they hope to be taken aboard by the rich lifeboat, since there are not as many people aboard and they could benefit from the supplies in that lifeboat.

  1. What should the passengers on the rich lifeboat do?
     a. Take the poor people aboard.
     b. Throw them some supplies such as food.
     c. Ignore them and let them drown.
     d. Shoot them and put them out of their suffering.

Let us now continue with our story and fill in some of the gaps you might have. First, you should understand that each lifeboat has a limit to its capacity to hold people. In other words, ecologists say it has a

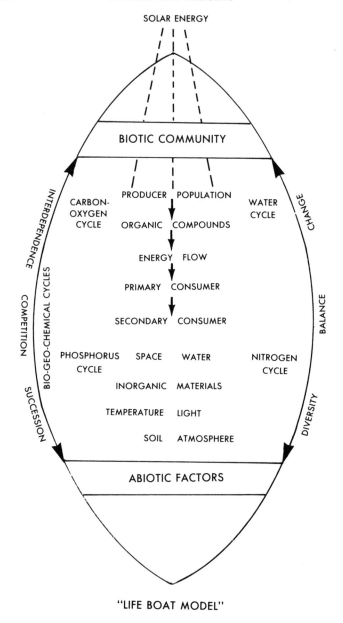

"LIFE BOAT MODEL"

carrying capacity based on the amount of food, medi-
cine, and other supplies for the number of people
aboard. For example, if we sit in a rich lifeboat that
has a carrying capacity of 50, no more could board
unless we are willing to give up the "quality of life"
we are accustomed to. To be generous, however, let
us assume our boat has a capacity of 10 more, making
60. This, however, is beyond the safety factor of the
lifeboat. For instance, if a change in the weather were
to take place, with 60 aboard all might drown, but
with only 50 they could survive.

1. What is the difference between a safety factor and
   carrying capacity?
2. How might you define "quality of life"?

Now let's say that 50 of us in the lifeboat see 100
others swimming in the water asking to be taken aboard
or to be given food. How would you respond to their
calls?

1. Take all 100 aboard, making the total number
   in the lifeboat 150.
2. Take only 10 of the 100, filling the lifeboat to a
   total of 60 passengers.
   a. How would you select the 10 to come aboard?
3. Take none aboard.
4. Change places with one in the water.

In studying our model, we have seen some of the

difficulties in making environmental decisions. This is exactly why all of us have these problems.

Let us now use our decision-making skills and facts to forecast or predict things that may become reality in the future.

Predictions of the Environmental Future

Below is a survey of various environmental issues. Study each carefully and check whether you approve or disapprove of its happening. After you have checked all of the predictions, go back and fill in the WHY boxes. See if you can give reasons for these predictions based on your understanding of this book and the facts you have gained from studying the environment.

Now that you have been involved in making predic-

Predictions	If it happens, I approve	I don't approve	WHY
1. Because of pollution, most drinking water will be supplied by desalination.			
2. Food farmed from the ocean will become one of the major sources of nutrition.			
3. The production and distribution of energy will be taken over and controlled by the government.			
4. Solar energy will be our major source of energy.			
5. People will be admired for buying and using up as little as possible of our natural resources.			

Predictions	If it happens,		WHY
	I approve	I don't approve	

6. Technology in the form of computers and automation will replace many people in factory and clerical jobs.

7. Many people will leave big cities and return to the farms.

8. American rivers and lakes will be free of pollution.

9. In many parts of the world, new cities will be planned and built for an uncontrolled population.

10. Automobiles will not be permitted in cities.

11. Heavy energy consumers such as powerboats and air conditioners will not be permitted.

12. Products will be designed to last a long time rather than wear out or go out of style.

13. Governments of all nations will agree to destroy all nuclear weapons.

14. There will be tax penalties to discourage couples from having more than two children.

15. Employed people will be required to retire at age 50.

16. A pollution-free automobile will be developed.

17. More efficient railroads will be developed to replace trucks.

18. Future nuclear power plants will be required to be underground.

19. Trash will be required to be sorted and re-mined for valuable resources.

20. The water will be cleaner in the U.S. because of strict enforcement of laws.

tions and giving reasons for them based on fact, it is time for you to go out and do something about YOUR environment. If you become actively involved in an environmental problem within your community, you can be considered a steward of your environment and can help the future of all environments on this planet.

Below are some activities that you can become involved in. Let's get together and save our earth; there is still time!

What Can I Do About the Environment?

Remember that environmental problems are a new issue for legislators and officials. They need information, and you may help to supply it. You may do this by forming groups or committees, since teamwork is important, or by developing a club or resource center in your community. Once this teamwork is established, you may become actively involved in activities such as the following:

1. Make sure your family car has a good muffler and pollution control devices. Maintain them.
2. Check your home heating system regularly to see that it operates efficiently.
3. Serve as a helper for a conservation nature club.
4. Develop an environmental center in your community by being the prime mover of such a center.

5. Work with city officials to set aside vacant lots for study areas and use by the community.

6. Raise money to finance the proper management of key natural areas or parks in your community.

7. Locate and publicize major areas of air, water, trash, and industrial pollution in your community.

8. Report water pollution sources to a pollution control agency and help them turn public sentiment toward correction.

9. Find out what laws and ordinances exist in your community and state for the prevention of water pollution. Proper enforcement of existing legislation can correct many environmental abuses.

10. Work through your local conservation group to establish a conservation commission as part of local government.

11. Examine existing zoning regulations in your community. Do they provide for present and future needs? If not, work with local groups to change them.

12. What does your community do about billboard advertising? Can it be eliminated or reduced?

13. Support industries that practice good conservation by purchasing their products.

14. Buy only returnable bottles; boycott disposable bottles and other disposable items.

15. Make posters that illustrate environmental problems and ask local merchants to display them.

16. Work with town officials and local merchants to plant trees and shrubbery along streets. Conduct surveys to determine what trees will grow on streets in your community.
17. Clean up litter in local parks.
18. Plant edible seed shrubs for wildlife feeding.
19. Plant trees and shrubs in parks and other areas to prevent erosion. Trees may be obtained free from some state conservation departments.

X

In Command of Tomorrow

Most of the environmental problems we face today result directly from increasing population and changing technology. Man demands more products that pollute water and air, makes more wastes than can be disposed of, uses biocides to increase food supply, and puts impossible demands on his ecosystem.

The first step toward developing an ecological awareness is to be informed. Just reading this book will give you a quick overview of the impacts man has made on his environment. But, in order to be better informed, you need to obtain environmental journals, read other books, and become knowledgeable about the issues.

To become knowledgeable in environmental issues takes special training and reading. You might even wish to make a career in this field. As you can see from the graph below, jobs in environmental fields are growing. In just five years' time the field has increased tremendously. Perhaps in the next five years it will jump even higher. As people become more and more aware of the importance of their environment, more and more jobs will open up to become stewards of the environment, teach about the environment, or do research in the environment.

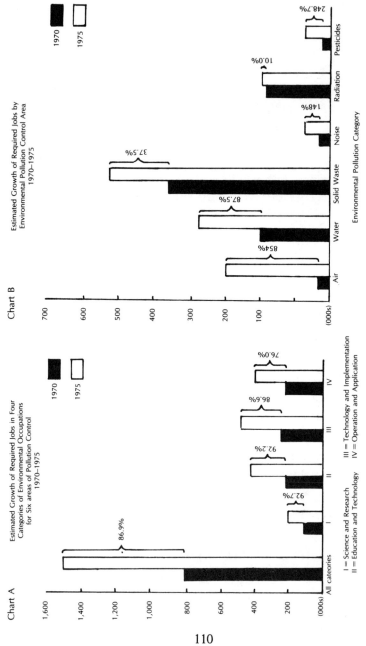

Chart A

Estimated Growth of Required Jobs in Four
Categories of Environmental Occupations
for Six areas of Pollution Control
1970–1975

1970
1975

I = Science and Research   III = Technology and Implementation
II = Education and Technology IV = Operation and Application

FIG. 13A

Chart B

Estimated Growth of Required Jobs by
Environmental Pollution Control Area
1970–1975

1970
1975

Environmental Pollution Category

FIG. 13B

110

The Grand Sweep: Neighborhood residents get out the "heavy artillery" to improve the appearance of their area.

You are the future of this planet, and only through your deeds and actions can this small planet and everything on it survive. To quote from Frances Thompson,

> All things by immortal power
> near or far
> Hiddenly

> To each other linked are,
> That thou canst not stir a flower
> Without troubling of a star.

Appendix

Environmental Careers

I. CAREERS IN ENVIRONMENTAL SCIENCE AND RESEARCH

Life Scientists
  General—Aquatic Biologist, Biochemist, Biophysicist, Biostatistician, Cytologist, Geneticist, Microbiologist, Pathologist; *Animal*—Animal Ecologist, Animal Husbandryman, Entomologist, Pharmacologist, Physiologist, Zoologist; *Plant*—Agriculturist, Agronomist, Botanist, Forest Ecologist, Horticulturist

Physical Scientists
  Astronomer; Chemist—Analytical, Inorganic, Organic; Geologist—Assayer, Geodist, Geophysicist, Hydrographer, Metallurgist, Meteorologist, Mineralogist; Oceanographer; Physicist; Seismologist; Soil Scientist

Social Scientists
  Anthropologist, Economist, Geographer, Mathematician, Political Scientist, Psychologist, Sociologist, Statistician, Writer

II. CAREERS IN ENVIRONMENTAL EDUCATION AND TECHNOLOGY

Environmental Educators

Camp Counselor; Humanities Teacher; Life Sciences Teacher—Biology Teacher, Physiology Teacher; Physical Sciences Teacher—Chemistry Teacher, Geography Teacher, Geology Teacher, Physics Teacher; Public Health Educator; Social Sciences Teacher; Vocational Teacher

Environmental Engineer

Aeronautical; Agricultural; Chemical; Civil; Combustion; Electrical; Environmental; Geological; Hydraulic; Industrial; Industrial Safety; Mechanical; Mining; Nuclear

Environmental Health

Dietitian and Nutritionist; Field Health Officer; Hygienist; Physician; Sanitarian; Toxicologist

Environmental Planners

Architect; Landscape Architect; Urban Planner

Natural Resource Managers

Fish and Game Warden; Forester; Oceanographer; Park Ranger—Naturalist; Range Manager; Soil Conservationist; Watershed Manager; Wildlife Manager

III. CAREERS IN ENVIRONMENTAL TECHNOLOGY AND IMPLEMENTATION

Inspectors or Monitors

Environmental Inspector, Food and Drug Inspector; Health Monitor; Nuclear Inspector

Technicians

Biological; Environmental; Food; Health; Horticultural; Land-Use; Nuclear; Physical Science; Resource Conservation

Testers or Analysts

Environmental; Mechanical

IV. CAREERS IN ENVIRONMENTAL APPLICATION AND OPERATION

Laborers—Attendants

Gardener; Incinerator Attendant; Janitor; Refuse Collector; Resource Developer; Wildlife Attendant

Operators and Foremen

Incinerator Foreman; Land-Fill Operator; Power Plant Operator; Recycling Operator; Water and Sewer System Foreman; Water Treatment Operator

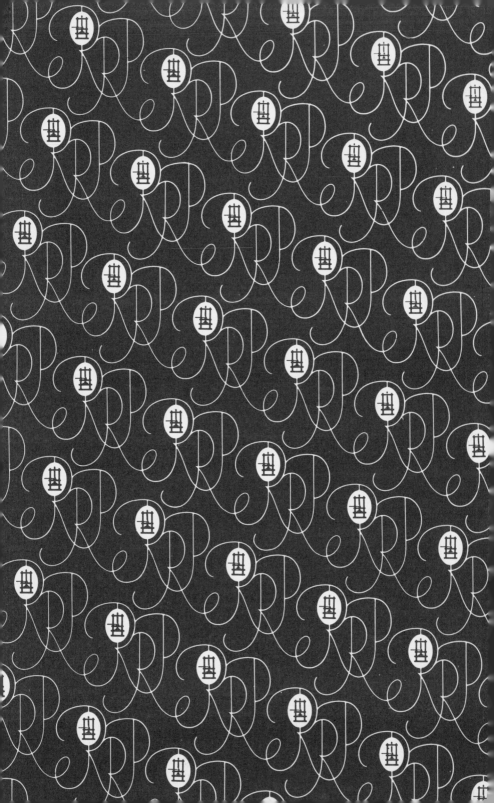